3/2015

Dr. Klaus Hoermann is a principal and partner at KUGLER MAAG CIE where he is responsible for the service area "Process Improvement" worldwide. He leads large and challenging process improvement projects and conducts assessments, appraisals, CMMI and SPICE trainings as well as assessor trainings and coaching. Dr. Hoermann is an iNTACS ISO/IEC 15504 Principal Assessor and trainer, SEI-authorized CMMI Instructor, and SCAMPI Lead Appraiser.

Markus Mueller
MAAG CIE. He is a qualified and experienced Project Management Professional, iNTACS ISO/IEC 15504 Principal Assessor and trainer. He serves also as an advisory board member of iNTACS. Mr. Mueller has been working in industry and research projects for more than 15 years, the last 8 years predominantly in the automotive industry, advising leading car manufacturers and suppliers on process improvements and assessments.

Lars Dittmann has been working as an EOQ auditor and software assessor for the Volkswagen Group in Wolfsburg since 1999. As the head of Software Quality Assurance he established QA measures such as the SPICE supplier assessments. At present he is responsible for IT Lifecycle and Quality within the IT department of the Volkswagen Group. He is an iNTACS ISO/IEC 15504 Principal Assessor and has conducted more than 40 assessments. Mr. Dittmann was the project leader of the Automotive SPICE® AUTOSIG Group and spokesman of the HIS working group "Process Assessment". He is also a member of the iNTACS Advisory Board.

Joerg Zimmer

charge of inter-divisional software quality projects. Furthermore, he is an iNTACS ISO/IEC 15504 Principal Assessor and a spokesman of the HIS Working Group "Process Assessment". Mr. Zimmer is also a member of the VDA Working Group 13 and a member of the Automotive SPICE® AUTOSIG Group.

Klaus Hoermann · Markus Mueller · Lars Dittmann · Joerg Zimmer

Automotive SPICE™ in Practice

Surviving Interpretation and Assessment

rockynook

Klaus Hoermann, klaus.hoermann@kuglermaag.com
Markus Mueller, markus.mueller@kuglermaag.com
Lars Dittmann, lars.dittmann@online.de
Joerg Zimmer, Joerg.Zimmer@daimlerchrysler.com

Editor: Jimi DeRouen

Copyeditor: James W. Johnson

Translator: Dieter Wachendorf

Layout and Type: Birgit Bäuerlein

Cover Design: Helmut Kraus, www.exclam.de

Printed in the USA

ISBN 978-1-933952-29-1

1st English Edition
© 2008 by Klaus Hoermann, Markus Mueller, Lars Dittmann, Joerg Zimmer

16 15 14 13 12 11 10 09 08   1 2 3 4 5

Rocky Nook Inc.
26 West Mission Street Ste 3
Santa Barbara, CA 93101

www.rockynook.com

First published under the title *Automotive SPICE in der Praxis: Interpretationshilfe für Anwender und Assessoren* © dpunkt.verlag GmbH, Heidelberg, Germany

Library of Congress Cataloging-in-Publication Data

Automotive SPICE in der Praxis. English
Automotive SPICE in practice : surviving interpretation and assessment / Markus Mueller ... [et al.].—1st ed.
    p. cm.
ISBN 978-1-933952-29-1 (alk. paper)
   1. Automobiles—Design and construction—Data processing. 2. Automobile industry and trade—Automation—Data procesing. 3. Automobile industry and trade—Management—Data processing. 4. Production management—Computer programs. 5. SPICE (Computer file) I. Mueller, Markus, 1968- II. Title.
TL278.A8835 2008
629.2'310285—dc22
                                               2008029002

Distributed by O'Reilly Media
1005 Gravenstein Highway North
Sebastopol, CA 95472

All product names and services identified throughout this book are trademarks or registered trademarks of their respective companies. They are used throughout this book in editorial fashion only and for the benefit of such companies. No such uses, or the use of any trade name, is intended to convey endorsement or other affiliation with the book.
No part of the material protected by this copyright notice may be reproduced or utilized in any form, electronic or mechanical, including photocopying, recording, or by any information storage and retrieval system, without written permission of the copyright owner.

This book is printed on acid-free paper.

Forword

by Dr. Karl-Thomas Neumann

The ability to develop software and electronics of the highest quality, within budget and a given time frame, is becoming a critical factor for today's success in a growing number of industries. This is a fact that the automotive industry, for instance, had to learn at a high price in recent years.

Today, companies depend on their ability to systematically identify, structure, and optimize their organizational and development processes. Many industries have come to recognize electronics and software expertise as an opportunity for innovation and as a key to their future. Consequently, possessing this know-how is becoming a core competence of their business. It is part of a cultural change that is taking place in organizations, where mechanical engineers are becoming mechatronic engineers, computer scientists are becoming quality controllers, and managers are implementing new insights and management practices. To master the complexity of electronics and software, even in commodity products, companies are forced to undertake completely new ways of development and to deploy new technologies. They must be able to operate a proactive risk management for software-related systems with ever-shorter product life cycles.

A good example is the positive influence on product quality via effectively implemented processes in development and production. This is why companies and their suppliers continuously improve their development processes. Process evaluations and 'best practice' models are providing a basis for this improvement. Global competition leads to new demands on functionality, quality, and reliability of products. Development times are getting shorter and development costs need to be reduced, while quality has to be ensured on a high level. Presently, no organization can afford a great number of unnecessary development loops. Cost drivers of a development process need to be identified and managed.

The DIN/EN/ISO 9001 certifications in the '90s have not led OEMs to have true confidence in the processes of their development partners. Thus, OEMs as well as suppliers use standards or defacto standards like ISO/IEC 15504 and CMMI today to effectively improve their development processes, to put projects on a solid footing, and to ensure the predictability of results. OEMs in industries like automotive, aerospace, and medical technologies also use ISO/IEC 15504 to identify reliable development partners as well as to evaluate the software development processes of their suppliers. Based on ISO/IEC 15504 it is possible to identify and control risks in development and procurement, and to systematically evaluate and advance the competency of development partners regarding software development. This provides a basis for lasting improvements.

Forword

Started in 1992, the SPICE project has been aiming to define a model for the evaluation of software development processes and to transfer this model into an ISO standard. Since then, the term SPICE itself has become a synonym for quality in development. With the publication of ISO/IEC 15504 industries now have an evaluation scheme at their disposal that was improved and advanced—in cooperation with various interested parties like the German automotive industry—based on the experiences with the Technical Report. The challenge consists in interpreting the standard correctly and in adapting it to the actual given problem. This currently available book, Automotive SPICE in Practice, offers the necessary interpretation aids and supports the reader to better understand the requirements of the standard in the context of a particular situation. It offers concrete examples from the development of software related systems (so-called »embedded systems«). Current trends regarding the further development of the standard are taken into consideration.

The authors are recognized experts with considerable expertise, which they obtained during more than one hundred assessments in the field. They have supported the implementation of improvement programs in numerous companies. I am confident that this book will further advance the understanding of the evaluation of maturity levels of software development, will be helpful in performing these evaluations, and that it will systematically support process improvements necessary in organizations.

Dr. Karl-Thomas Neumann
Member of the Executive Board—CTO Continental Group

Foreword

by Alec Dorling

In today's environment, when modern vehicle manufacturers are improving their reputations for performance, quality, and value for money, succeeding in the market place is ever more dependent on the ability to employ new technology in all aspects of the business. At the same time there is market demand to deliver products with less cost and reduced timescales.

Increasing competition creates strong pressures for even more features, quality, and reliability while at the same time demanding delivery with less cost and reduced timescales. As cost and functionality increase so does the complexity, bringing with it additional problems such as subsystem interoperability, in-service reliability, and predictability of project development times.

It is widely understood that the quality of the end product can largely be determined by the quality of the processes that produce it. There is general consensus that the better the process, the more accurate are the plans, the better is the product quality, and the earlier are the deliverables.

Traditional reliance on an organization having certification to a recognized Quality System Standard such as ISO9001 has not provided sufficient confidence in the software area, especially concerning suppliers of safety-critical embedded software systems.

The concepts of *process* and *maturity* are increasingly being applied both as a framework for assessment and for improvement. Focusing on specific process improvement objectives allows an organization to conduct progressive improvement and to build mature processes to meet the needs of the business and to adapt to changes in the business environment.

Customers as well are providing an increasing focus on the capability and maturity of the supplier in the supplier evaluation process, making provision for contractual demands for required process capability levels with reference to the ISO/IEC 15504 Standard for Process Assessment as a means to identify and control risk, both before and during contract performance.

With the publication of the ISO/IEC 15504 Standard for Process Assessment during 2003–2006 (originally published in 1998), industry now has a recognized standard that provides a solid and mature framework for performing assessments. The international standard provides a common framework for assessment with a range of process assessment models covering systems, software, and service that embody a set of best practices that can be used for any process improvements.

The SPICE project was established as an international project to initiate the development of the standard and to support trials and adoption of the standard.

The SPICE User Group with its member community has been organizing the SPICE international conferences and workshops around the world. 'SPICE', however, has become synonymous with the ISO/IEC 15504 standard itself to the community at large.

The SPICE User Group and The Procurement Forum have been working with the major motor vehicle manufacturers, through the Automotive SPICE© initiative, to address the specific needs of the automotive sector and to develop a common framework for the assessment of suppliers in the automotive industry based on ISO/IEC 15504.

The culmination of this work has resulted in the publication of the Automotive SPICE© Process Assessment Model which is based on the published ISO/IEC 15504 Exemplar Process Assessment Model, but is further tailored for use and supplemented by automotive guidance for application in the automotive domain.

While the Automotive SPICE© Process Assessment Model does provide guidance for automotive application, it cannot in itself provide the detailed insight and subtle nuances in interpretation that can only be provided by knowledgeable and experienced assessors.

Automotive SPICE in Practice: Surviving Implementation and Assessment provides that much needed guidance for users, companies, and assessors in the interpretation and application of the standard and the Automotive SPICE© Process Assessment Model for both process improvement and process capability determination.

The authors are recognized professionals with significant knowledge and experience, having jointly performed more than 100 assessments in the field and having helped many companies with their improvement programs. The importance of using qualified and experienced assessors in performing assessments should not be underestimated.

Throughout the book, the authors provide an insight into the standard and the Automotive SPICE© Process Assessment Model, providing interpretation of the text and elaboration of the work products. The many tips on how to apply the assessment process and on the success factors in planning and implementing improvements are invaluable. Templates are also provided to support implementation supplementing the guidance and best practices.

I hope that by reading this book you will gain useful insight into the Automotive SPICE© Process Assessment Model and that it will assist you in implementing process assessments and improvements in your organizations.

Alec Dorling
Impronova AB
Convener of the ISO/IEC 15504 international standardization group
Coordinator of the Automotive SPICE© initiative
President of the SPICE Academy

Preface

In our many years of work with models like SPICE, Automotive SPICE, and CMMI, we have repeatedly noticed how difficult it is to understand the requirements of these models, and how differently they are sometimes interpreted. Of course, since they are intended to suit a broad range of applications, it is in the nature of such models to have a wide interpretational scope. They include practices and work products that must be interpreted in their respective application context, ranging from a project scope to an entire organization. The same applies to the question regarding how model elements are to be evaluated in assessments. There are, therefore, no absolute standards regarding the fulfillment of the models' requirements.

Since the publication of our book »SPICE in der Praxis[1]« (dpunkt.verlag) in 2006, the automotive industry has completed the transition from SPICE to Automotive SPICE. We think that now is the right time to provide interpretational assistance for Automotive SPICE, in the hope that readers will find it helpful during assessments and process implementations.

Structure of the book:

- **Chapter 1** (Introduction and Overview) provides an introduction into the fundamental concepts of maturity models and a brief summary of their history, interrelationships, and trends. Moreover, it conveys, in a concise and comprehensible way, the basic knowledge required for a proper understanding of the structure and elements of Automotive SPICE.
- **Chapter 2** (Interpretations Regarding Process Dimension) examines a practice-oriented selection of Automotive SPICE processes in detail, explaining the purpose, base practices, and work products of each process. The crucial topic of »traceability« is dealt with in a separate section.

---

1. Published in German only by dpunkt.verlag, Heidelberg, 2006.

- **Chapter 3** (Interpreting the Capability Dimension) explains how capability levels are rated, providing a detailed description and interpretation of the capability levels, process attributes, and generic practices. The chapter also includes many notes providing additional guidance for assessors.
- **Chapter 4** (CMMI – Differences and Similarities) takes into account the fact that, in the automotive industry, CMMI is often applied in addition to Automotive SPICE. It compares structures, contents, and assessment methods of the two models, thus providing a first orientation-guide.
- **Chapter 5** (Functional Safety) provides a brief overview of IEC 61508, which will undergo a larger rollout in the automotive industry within the next few years. IEC 61508 also has requirements on development processes and methods. Many of the Automotive SPICE processes contribute to the accomplishment of these requirements, although, in many cases, IEC-61508 requirements exceed Automotive SPICE requirements by far.
- The **appendix** contains a selection of work products, a glossary, a list of abbreviations, web pages, literature, and standards.

All Automotive SPICE model texts are shown in italics. We decided to print the original passages unchanged although some were found to be grammatically incorrect and difficult to understand. Whereever necessary, we have added explanatory footnotes.

Our interpretational guide is based on practical experience which we recorded with utmost care. Nevertheless, we must point out that evaluating compliance with Automotive SPICE always requires the individual assessment of an organization or project. We do not wish to change this situation in any way, nor do we give any guarantee for the success of implementations based on interpretations, recommendations, and examples provided in this book.

We would like to express our thanks to our translator, Mr. Dieter Wachendorf (KUGLER MAAG CIE), who did an excellent job in translating this difficult subject matter, and to Mrs. Christa Preisendanz (dpunkt.verlag) for her professional support. We also thank our families for their continuous and patient understanding when we needed time away to complete this book.

Klaus Hoermann, Markus Mueller, Lars Dittmann, Joerg Zimmer
March 2008

Table of Contents

1	**Introduction and Overview**		**1**
1.1	Introducing the Subject Matter		1
1.2	Automotive SPICE and Other Maturity Models: History, Background, and Trends		2
1.3	Automotive SPICE: Structure and Components		6
	1.3.1	The Process Dimension	7
	1.3.2	The Capability Dimension	8
2	**Interpretations Regarding the Process Dimension**		**11**
2.1	ACQ.4 Supplier Monitoring		16
	2.1.1	Purpose	16
	2.1.2	Characteristics Particular to the Automotive Industry	16
	2.1.3	Base Practices	17
		Experience Report	20
		Experience Report	21
	2.1.4	Designated Work Products	22
	2.1.5	Characteristics of Level 2	23
2.2	SPL.2 Product Release		23
	2.2.1	Purpose	23
	2.2.2	Characteristics Particular to the Automotive Industry	24
	2.2.3	Base Practices	25
	2.2.4	Designated Work Products	29
	2.2.5	Characteristics of Level 2	30

2.3	ENG.1 Requirements Elicitation	30
	2.3.1 Purpose	30
	2.3.2 Characteristics Particular to the Automotive Industry	32
	2.3.3 Base Practices	32
	2.3.4 Designated Work Products	37
	2.3.5 Characteristics of Level 2	38
2.4	ENG.2 System Requirements Analysis	38
	2.4.1 Purpose	38
	2.4.2 Characteristics Particular to the Automotive Industry	38
	Excursus: System	39
	2.4.3 Base Practices	40
	2.4.4 Designated Work Products	45
	2.4.5 Characteristics of Level 2	46
2.5	ENG.3 System Architectural Design	46
	2.5.1 Purpose	46
	2.5.2 Characteristics Particular to the Automotive Industry	47
	2.5.3 Base Practices	48
	2.5.4 Designated Work Products	51
	2.5.5 Characteristics of Level 2	51
2.6	ENG.4 Software Requirements Analysis	52
	2.6.1 Purpose	52
	2.6.2 Characteristics Particular to the Automotive Industry	52
	2.6.3 Base Practices	53
	Excursus: Example Method Hazard and Operability Study (HAZOP)	54
	2.6.4 Designated Work Products	57
	2.6.5 Characteristics of Level 2	57
2.7	ENG.5 Software Design	57
	2.7.1 Purpose	57
	2.7.2 Characteristics Particular to the Automotive Industry	58
	2.7.3 Base Practices	58
	2.7.4 Designated Work Products	63
	2.7.5 Characteristics of Level 2	65

Table of Contents

2.8 ENG.6 Software Construction 65
    2.8.1 Purpose 65
    2.8.2 Characteristics Particular to the Automotive Industry 65
    2.8.3 Base Practices 66
    2.8.4 Designated Work Products 73
    2.8.5 Characteristics of Level 2 73

    Excursus: Test Documentation According to IEEE-Standard 829-1998 (Software Test Documentation) 74

2.9 ENG.7 Software Integration Test 75
    2.9.1 Purpose 75
    2.9.2 Characteristics Particular to the Automotive Industry 75
    2.9.3 Base Practices 76
    2.9.4 Designated Work Products 84
    2.9.5 Characteristics of Level 2 85

2.10 ENG.8 Software Test 85
    2.10.1 Purpose 85
    2.10.2 Characteristics Particular to the Automotive Industry 86
    2.10.3 Base Practices 87

    Excursus: A Brief Overview of Test Methods 89

    Excursus: Some Methods for the Derivation of Test Cases 89

    2.10.4 Designated Work Products 90
    2.10.5 Characteristics of Level 2 90

2.11 ENG.9 System Integration Test 91
    2.11.1 Purpose 91
    2.11.2 Characteristics Particular to the Automotive Industry 91
    2.11.3 Base Practices 92
    2.11.4 Designated Work Products 95
    2.11.5 Characteristics of Level 2 95

2.12 ENG.10 System Testing 96
    2.12.1 Purpose 96
    2.12.2 Characteristics Particular to the Automotive Industry 96
    2.12.3 Base Practices 97
    2.12.4 Designated Work Products 99
    2.12.5 Characteristics of Level 2 99

2.13	SUP.1 Quality Assurance	99
	2.13.1 Purpose	99
	2.13.2 Characteristics Particular to the Automotive Industry	100
	2.13.3 Base Practices	101
	2.13.4 Designated Work Products	110
	2.13.5 Characteristics of Level 2	112
2.14	SUP.2 Verification	113
	2.14.1 Purpose	113
	2.14.2 Characteristics Particular to the Automotive Industry	114
	2.14.3 Base Practices	114
	2.14.4 Designated Work Products	118
	2.14.5 Characteristics of Level 2	119
2.15	SUP.4 Joint Review	119
	2.15.1 Purpose	119
	2.15.2 Characteristics Particular to the Automotive Industry	121
	2.15.3 Base Practices	121
	2.15.4 Designated Work Products	125
	2.15.5 Characteristics of Level 2	126
2.16	SUP.8 Configuration Management	127
	2.16.1 Purpose	127
	2.16.2 Characteristics Particular to the Automotive Industry	127
	2.16.3 Base Practices	128
	2.16.4 Designated Work Products	135
	2.16.5 Characteristics of Level 2	137
2.17	SUP.9 Problem Resolution Management	137
	2.17.1 Purpose	137
	2.17.2 Characteristics Particular to the Automotive Industry	138
	2.17.3 Base Practices	138
	2.17.4 Designated Work Products	147
	2.17.5 Characteristics of Level 2	148
2.18	SUP.10 Change Request Management	149
	2.18.1 Purpose	149
	2.18.2 Characteristics Particular to the Automotive Industry	150
	2.18.3 Base Practices	151
	2.18.4 Designated Work Products	155
	2.18.5 Characteristics of Level 2	156

2.19	MAN.3 Project Management		156
	2.19.1	Purpose	156
	2.19.2	Characteristics Particular to the Automotive Industry	157
	2.19.3	Base Practices	157
	2.19.4	Designated Work Products	169
	2.19.5	Characteristics of Level 2	172
2.20	MAN.5 Risk Management		173
	2.20.1	Purpose	173
	2.20.2	Characteristics Particular to the Automotive Industry	174
	2.20.3	Base Practices	174
	2.20.4	Designated Work Products	180
	2.20.5	Characteristics of Level 2	181
2.21	MAN.6 Measurement		182
	2.21.1	Purpose	182
		Excursus: Goal/Question/Metric (GQM)Method	183
	2.21.2	Characteristics Particular to the Automotive Industry	184
	2.21.3	Base Practices	184
	2.21.4	Designated Work Products	190
	2.21.5	Characteristics of Level 2	193
2.22	PIM.3 Process Improvement		193
	2.22.1	Purpose	193
	2.22.2	Characteristics Particular to the Automotive Industry	194
	2.22.3	Base Practices	194
	2.22.4	Designated Work Products	199
	2.22.5	Characteristics of Level 1-3	199
2.23	REU.2 Reuse Program Management		200
	2.23.1	Purpose	200
	2.23.2	Characteristics Particular to the Automotive Industry	200
	2.23.3	Base Practices	201
	2.23.4	Designated Work Products	204
	2.23.5	Characteristics of Levels 1-3	205
2.24	Traceability in Automotive SPICE		205
	2.24.1	Introduction	205
	2.24.2	Key-Notes	205
		Excursus: Verification Criteria	208

3 Interpreting the Capability Dimension — 213

- 3.1 The Structure of the Capability Dimension 213
  - 3.1.1 Capability Levels and Process Attributes 213
  - 3.1.2 Process Capability Indicators 214
- 3.2 How Are Capability Levels Measured? 214
- 3.3 The Capability Levels 216
  - 3.3.1 Level 0 (»Incomplete Process«) 216
  - 3.3.2 Level 1 (»Performed Process«) 217
  - 3.3.3 Level 2 (»Managed Process«) 219
  - 3.3.4 Level 3 (»Established Process«) 234
  - 3.3.5 Level 4 (»Predictable Process«) 249
  - 3.3.6 Level 5 (»Optimizing Process«) 250

4 CMMI – Differences and Similarities — 253

- 4.1 Introduction 253
- 4.2 Comparison of Structures 255
- 4.3 Comparison of Contents 257
- 4.4 Comparison of the Assessment/Appraisal Methods 260

5 Functional Safety — 263

A Overview of Selected Work Products — 267

Glossary — 269

Abbreviations — 281

Literature, Standards, and Web Pages — 283

Index — 289

1 Introduction and Overview

This chapter consists of three parts: Section 1.1 deals with the question of why there is an increasing trend in the application of maturity models. Section 1.2 briefly explores the history of the maturity models, in particular with regard to the competing models CMMI®[1], SPICE, and Automotive SPICE™[2], and emerging trends. Section 1.3 explains the basic structures of Automotive SPICE that are needed to understand subsequent chapters.

1.1 Introducing the Subject Matter

In today's globalized world economy, products and services are rarely developed in isolation by a single company. Manufacturers are increasingly forced to develop their products within a whole network of in-house development centers, suppliers, and equal partners. The key driver is continually increasing costs, which forces companies to shift development to low cost locations and to engage in strategic partnerships. Since products and services are becoming increasingly complex and sophisticated while development cycles are getting shorter and shorter, two major issues have emerged:

- How can these complex cooperations and value chains be controlled?
- What can be done to ensure quality and adherence to cost and schedules?

For many businesses this has become a major challenge with an immediate effect on market success and growth. Key success factors relating to these questions are systematic and controllable business processes, especially those relating to management, engineering, quality assurance, acquisition, and cooperation with external partners. The methodology underlying the »maturity models« lends itself ideally to tackle these problems.

---

1. CMMI is registered in the U.S. Patent and Trademark Office.
2. »Automotive SPICE« is a registered trademark of Volkswagen AG, Wolfsburg.

Maturity models such as SPICE, Automotive SPICE, and CMMI, offer suitable methods (by means of tried »good practices«); they have a strong focus on processes[3] and have, for many years, been succesfully applied to address the issues stated above. In the Nineties, the CMM®[4] and later CMMI models became a real success in the U.S. The initial push came from the US Department of Defense which suffered from enormous cost and budget overruns caused by its software and system suppliers and achieved significant improvements applying CMM/CMMI. Today, it is common practice in the U.S. to require a certain CMM or CMMI level from suppliers and certification through audits[5]. No bid will be accepted without such a level. A similar situation has emerged in the car industry with regard to Automotive SPICE.

1.2 Automotive SPICE and Other Maturity Models: History, Background, and Trends

The idea of the maturity models was also taken up in Europe. Particularly SPICE and, more recently, Automotive SPICE are being applied there, in addition to CMM and CMMI.

Figure 1–1 shows the most important maturity models and their historical development. The oldest model is CMM (see [CMM 1993a], [CMM 1993b]), which used to be very extensively deployed internationally but which has now been superseded by CMMI. In the car industry, CMM has never played an important role, although it was used by one car manufacturer for a short period to try out approaches to supplier evaluation. A few pioneers among the suppliers to the car industry used BOOTSTRAP, a model compatible to SPICE that never really caught on as an alternative to SPICE and was discontinued in 2003.

SPICE (see [Hoermann et al. 2006]) evolved from an ISO[6] project of the same name and was published in 1998 as ISO/IEC TR 15504, whereby TR (Technical Report) is a forerunner to a later International Standard (IS). The different parts of the International Standard ISO/IEC 15504 were successively published from 2003 onwards. Part 5, which from a practical point of view is the most important part, was issued in 2006.

---

3. It has been a proven principle for some decades that good and controlled processes (in addition to qualified staff and command of the technology) have a demonstrably positive influence on quality and cost.
4. The term CMM is registered in the U.S. patent and Trademark Office.
5. So-called »assessments«, »evaluations« and »appraisals«.
6. International Organization for Standardization.

1.2 Automotive SPICE and Other Maturity Models

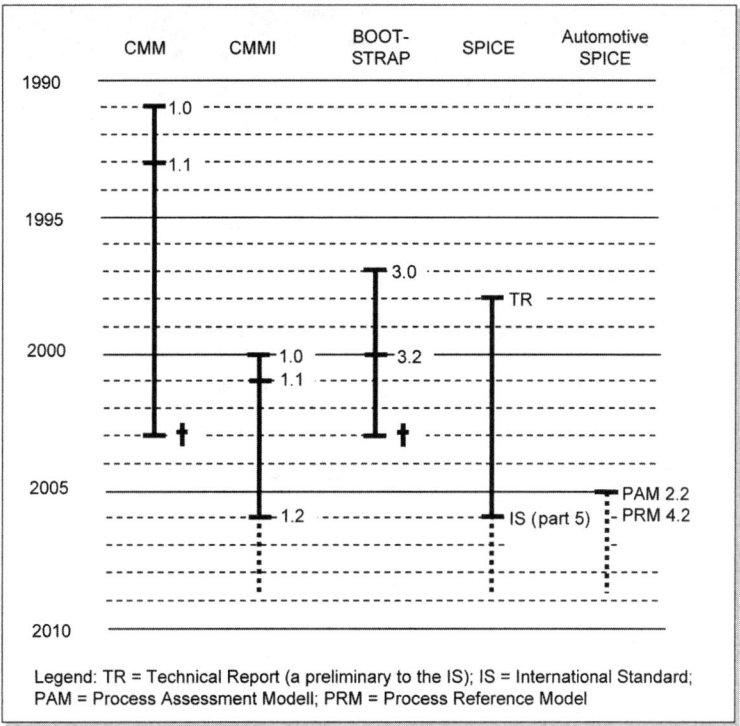

Figure 1-1 History of the most important maturity models used in Europe (only published models that have been applied in practice)

The breakthrough in the use of maturity models in the car industry came in 2001 with the decision of the OEM software initiative (HIS[7]) to use SPICE for the evaluation of suppliers in the software and electronics sector. Since then the application of SPICE has spread throughout the car industry. One of the big advantages of SPICE is that it allows the creation of domain-specific models consolidated under one common umbrella. Besides the space industry, the car industry also made use of this feature: In 2005 the Automotive Special Interest Group[8] (AUTOSIG) of the Procurement Forum published the Automotive SPICE-Model (see [Automotive SPICE]). This model is now being applied by participating car manufacturers to perform assessments of their software and electronics suppliers. Figure 1-2 shows how the processes required by the HIS group, based on ISO/IEC TR 15504, correspond to the new requirement based on Automotive SPICE.

---

7. The working group established by Audi, BMW, Daimler (previously DaimlerCrysler), Porsche and Volkswagen.
8. In addition to Audi, BMW, Daimler (previously DaimlerCrysler), Porsche and Volkswagen, other participating car manufacturers are Fiat Auto, Volvo Car Corporation (together with Ford Europe, Jaguar, and Land Rover).

1 Introduction and Overview

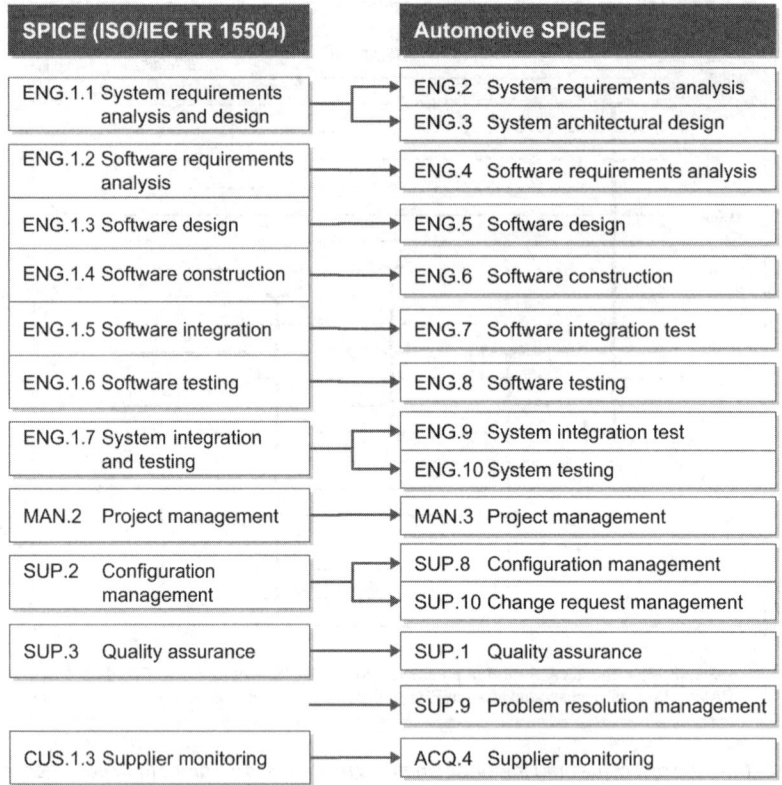

Figure 1–2   *Mapping of processes required by the HIS working group based on ISO/IEC TR 15504 processes to the new requirement, based on Automotive SPICE (no mapping for problem resolution management)*

Besides SPICE and Automotive SPICE, CMMI has also become increasingly prevalent in the motor industry (see [Kneuper 2006, Ahren et al. 2001, Chrissis et al. 2003, CMMI 2006]). In fact, in a survey of the »Hansen Report on Automotive Electronics« [Hansen], CMMI is mentioned more often than SPICE (see figure 1–3). The data shown refers to internal application, i.e., in-house development. With respect to suppliers, SPICE and Automotive SPICE are usually requested. The reason for CMMI's wide-spread market penetration is due to the advantages of the model[9], its good, world-wide support (see [SEI], [SEI Rep]) and the fact that its rapid growth since 2000 has made it evolve into a kind of quasi-standard. Even if there are no official figures available for SPICE and its derivatives we may assume the figure for CMMI to be considerably higher: by April 2007, approx. 61,000 people had attended the official three-day CMMI introductory course world-wide, and the figures show a sharply increasing trend.

---

9.   Examples are the models' degree of detail, its comprehensibility, and generic applicability for software, hardware, systems, and services alike.

1.2 Automotive SPICE and Other Maturity Models

Many organizations require expertise in several models:

- Vehicle manufacturers require Automotive SPICE accreditation from their suppliers, but if they are presented with CMMI appraisal results because supplier processes are geared to comply with CMMI, then manufacturers must be in a position to evaluate them.
- Suppliers whose processes are designed primarily to comply with the CMMI model must design their processes to satisfy the requirements of other models, too.

In the next section we are going to discuss the differences between SPICE and Automotive SPICE, whereas in chapter 4 we shall consider CMMI, SPICE, and Automotive SPICE.

From an application point of view, in particular, the emerging trends are of interest. Currently known facts are:

- SPICE: Work is in progress for the following new parts of the standard:
  - ISO/IEC TR 15504 Part 6—Exemplar Systems Life Cycle Processes Assessment Model—Working Draft
  - ISO/IEC 15504 Part 7—Assessment of Organizational Maturity—Working Draft
  - Official and reliable due dates are not yet available.
- Automotive SPICE: Since the release of PAM 2.3[10], no further development has been announced.
- With version 1.2, CMMI received a new architecture that combines domain specific models (called »constellations«) by means of commonly used model components. The CMMI constellations for »Development« and »Acquisition« were released in 2006 and 2007, respectively, and in 2008 the constellation for »Services« is supposed to be available. For many organizations this may be very interesting from a strategic point of view, as this will make it possible to standardize principal processes of different business divisions (development, IT/services, purchasing).

A possible trend can be recognized: the expansion from the successful application in software development to further domains, in particular to system development, mechanics, and hardware development. It stands to reason to apply in other development areas what has been tried and tested in the software domain. Whereas SPICE is currently preparing to expand towards system development with its new part 6, CMMI has already completed this move (software, hardware, systems, services). Whether and in which way the HIS initiative or the Automotive Specific Interest Group (AutoSIG) will follow this trend is unknown but not unlikely. After all, quality problems are not only caused by software components. Some companies, such as Bosch (see figure 1–3) have obviously recognized this trend very early and are already proactively extending in the direction

---

10. In this book we refer to PAM 2.3.

towards hardware and mechanics development. Another trend is that organizations are moving to higher levels. Many enterprises, especially in Asia, did this with CMMI, thereby demonstrating their ability to develop high quality software efficiently. Now this trend is also beginning with Automotive SPICE. One example is Continental Automotive Singapore, Pte Ltd who recently achieved Automotive SPICE Level 4 in several processes.

	Organization	Favored Process
BMW	E/E Process Chain	CMMI in addition to specific BMW targets
Chrysler	E/E Core Engineering	CMMI/SPICE
General Motors	Powertrain, Vehicle Engineering	GM-specific, based on parts of CMMI
Honda	E/E Systems R&D	Considering CMMI and other technologies
Mercedes Car Group	USA Germany	CMMI SPICE
Toyota		Proprietary, based somewhat on CMMI
Volkswagen	E/E Engineering R&D	SPICE
Bosch Automotive	All software business units, plus component development departments	CMMI/SPICE
Continental Automotive Systems	All business units that deliver software	SPICE
Delphi	Electronics & Safety	CMMI
Siemens VDO	All 13 divisions	CMMI
Valeo	All devisions that deliver software	CMMI/SPICE
Visteon		Visteon Engineering Process (VEP), which uses elements of CMMI and SPICE

Figure 1–3   *Distribution of maturity models in the car industry (excerpt) source: [Hansen], July 2005, p. 8*

How will the models' market shares develop? SPICE will probably continue to lose ground as a practical application, due to the motor industry's turn towards Automotive SPICE, while the latter will become firmly established in the supplier evaluation domain. In many organizations, CMMI will probably continue to gain ground and be used as an internally applied instrument, especially with regard to the strategic aspects described earlier.

1.3   Automotive SPICE: Structure and Components

Automotive SPICE consists of two components, a **process reference model (PRM)** and a **process assessment model (PAM)**. For practical use it is sufficient to be familiar with the process assessment model; therefore we shall confine ourselves here to a discussion of the process assessment model.[11]

---

11. See [Hoermann et al. 2006] for a detailed discussion of the wider context.

1.3 Automotive SPICE: Structure and Components

The process assessment model contains all the details needed to evaluate process capability (so-called indicators) and is organized in two dimensions:

- Process dimension:
  The process dimension contains indicators for all processes to evaluate the extent to which these processes are performed. These indicators differ from process to process and form an important prerequisite for the achievement of level 1 (described in section 1.3.2).

- Capability dimension:
  The capability dimension contains the indicators for the evaluation of the different capability levels. These indicators are identical for all processes.

1.3.1 The Process Dimension

Automotive SPICE processes are shown in figure 1–4.

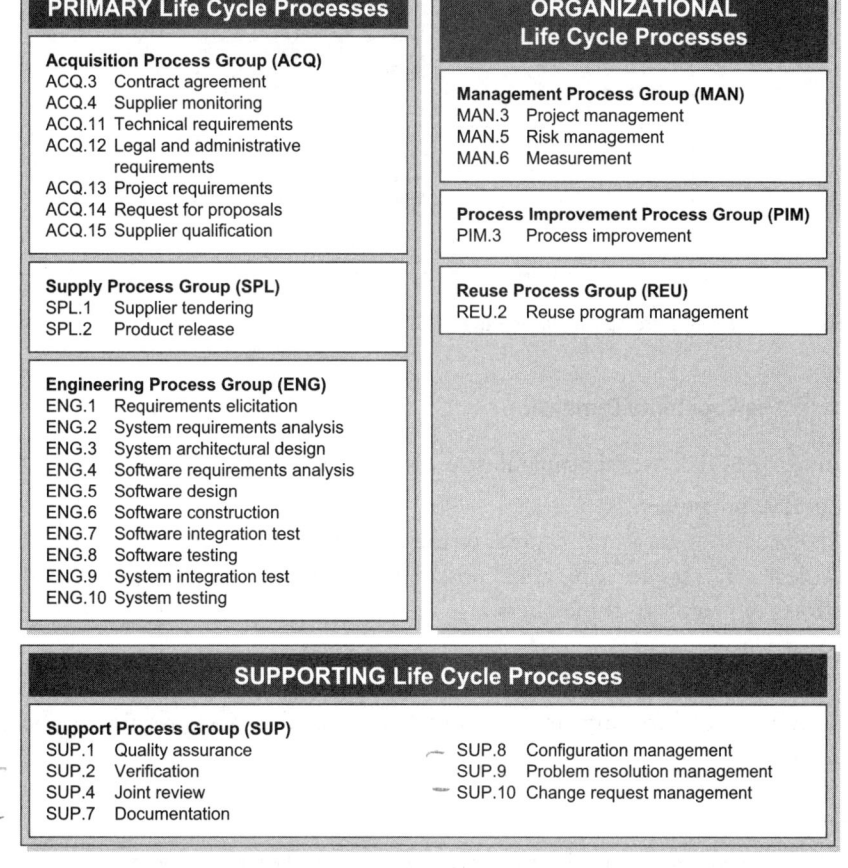

Figure 1–4   *Processes in Automotive SPICE*

Each process is structured in the same way, as shown in figure 1–5. Base practices are exemplary activities[12] needed to accomplish the process outcomes. Process outcomes are the detailed results of the achieved process purpose specifying what is to be accomplished by the process. Output work products are exemplary, typical results of a process; however, they are not mandatory. Together with the base practices they are the objective proof that the process has achieved its purposes. For that reason they are called process performance indicators and constitute the criteria for the achievement of Level 1.

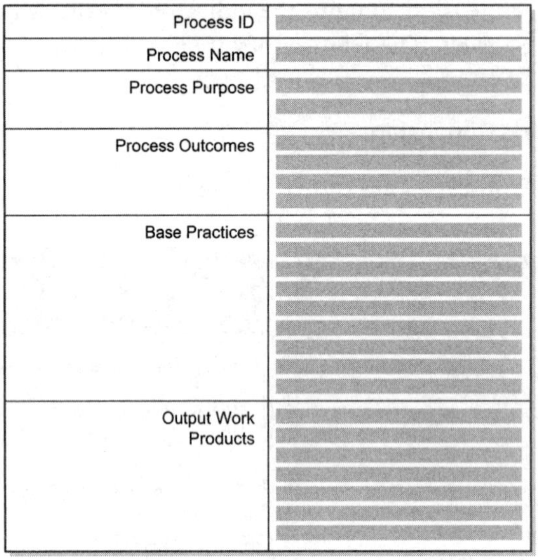

Figure 1–5   Process structure in Automotive SPICE

1.3.2 The Capability Dimension

Automotive SPICE uses six capability levels for processes (see figure 1–6):

- **Level 0:** *Incomplete*
  The process is not implemented, or the purpose of the process is not fulfilled. Project successes are a practical possibility but based solely on the individual efforts of project staff members.
- **Level 1:** *Performed*
  The implemented process fulfills the purpose of the process. This means that basic practices are implemented and that defined process results are being achieved.

---

12. All Automotive SPICE model elements are exemplary, i.e., they specify, »what« needs to be implemented, but not »how«. They are requirements on an organization's processes. The processes of an organization may, however, be named and structured quite differently.

1.3 Automotive SPICE: Structure and Components

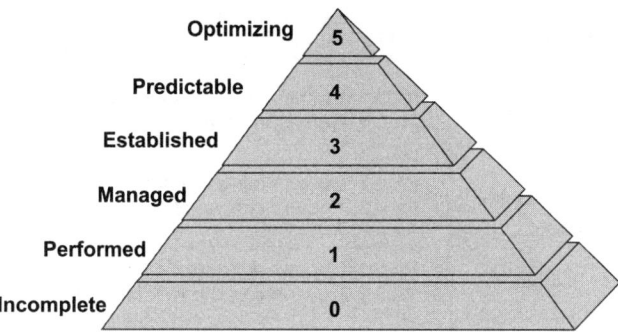

Figure 1–6   *The six capability levels*

- Level 2: *Managed*
  Process performance is now additionally planned and tracked, and planning is continuously adjusted. The work products of the processes are adequately implemented; they are under configuration management, quality assured, managed, and adjusted.

- Level 3: *Established*
  A standard process is now established and valid throughout the organization. Projects use an adapted version of this standard process (a so-called »defined process«) that is derived by means of »tailoring«. This process is able to achieve defined process results.

- Level 4: *Predictable*
  While performing the defined process, detailed measurements are performed and analyzed, leading to a quantitative understanding of the process and improved predictability. Statistical methods are applied to control the defined process between upper and lower limits. The quality of work products is quantitatively known.

- Level 5: *Optimizing*
  Quantitative process objectives are defined based on the organizations's business goals, and are permanently tracked. Processes are continuously improved, innovative approaches and techniques are tried and replace less effective processes to better achieve predefined objectives.

Whether a particular process capability level is achieved is determined by means of process attributes. Process attributes are assigned to capability levels and characterize them as regards content (see figure 1–7). Each process attribute defines a particular aspect of process capability. Level 2, for instance, is defined by the attributes »performance management« (i.e., planning, assignment of responsibilities and allocation of resources, monitoring etc.) and »work product management« (i.e., ensuring that the requirements on work products are fulfilled).

Capability level	Process attributes
5 Optimizing	PA 5.1 Process innovation attribute PA 5.2 Process optimization attribute
4 Predictable	PA 4.1 Process measurement attribute PA 4.2 Process control attribute
3 Established	PA 3.1 Process definition attribute PA 3.2 Process deployment attribute
2 Managed	PA 2.1 Performance management attribute PA 2.2 Work product management attribute
1 Performed	PA 1.1 Process performance attribute
0 Incomplete	

Figure 1-7 The process attributes

Details of process attributes and their evaluation are discussed in chapter 3. Processes are evaluated according to a four-point **rating scale**:

- N Not achieved
- P Partially achieved
- L Largely achieved
- F Fully achieved

The process capability level is calculated based on the process attribute evaluations following a simple calculation rule (see chapter 3). In order to reach a particular capability level, the rating of process attributes of that level must be at least L, and all process attributes of the lower capability levels must be rated F.

2 Interpretations Regarding the Process Dimension

It would go beyond the scope of this book to try and cover all the processes of Automotive SPICE; this is why we must confine ourselves to a useful selection. We decided to concentrate on processes which, based on our experience from improvement projects and many assessments, were typically of central relevance for most users—at least at the beginning of their improvement activities. Furthermore, we wanted to cover at least those processes that are evaluated by the OEM[1] software initiative (HIS) (see chapter 1) in the context of supplier assessments, as they constitute by far the largest volume of assessments. The results are shown in figure 2–1, and processes covered in this book are marked with a bullet.

Every process in Automotive SPICE consists of a standardized structure (see figure 1–5) and a set of individual process elements:

Process ID: Unique process identifier, consisting of a combination of three letters, a period, and one number between 1 and 12 (e.g., »MAN.3«).

Process Name

Process Purpose

Process Outcomes: The defined process outcomes

Base Practices: Practices constituting essential functional elements of the process.

Output Work Products: Each work product has a unique WP-ID and is explained in Automotive SPICE PAM, Annex B (work product characteristics).

---

1. (Original Equipment Manufacturer)

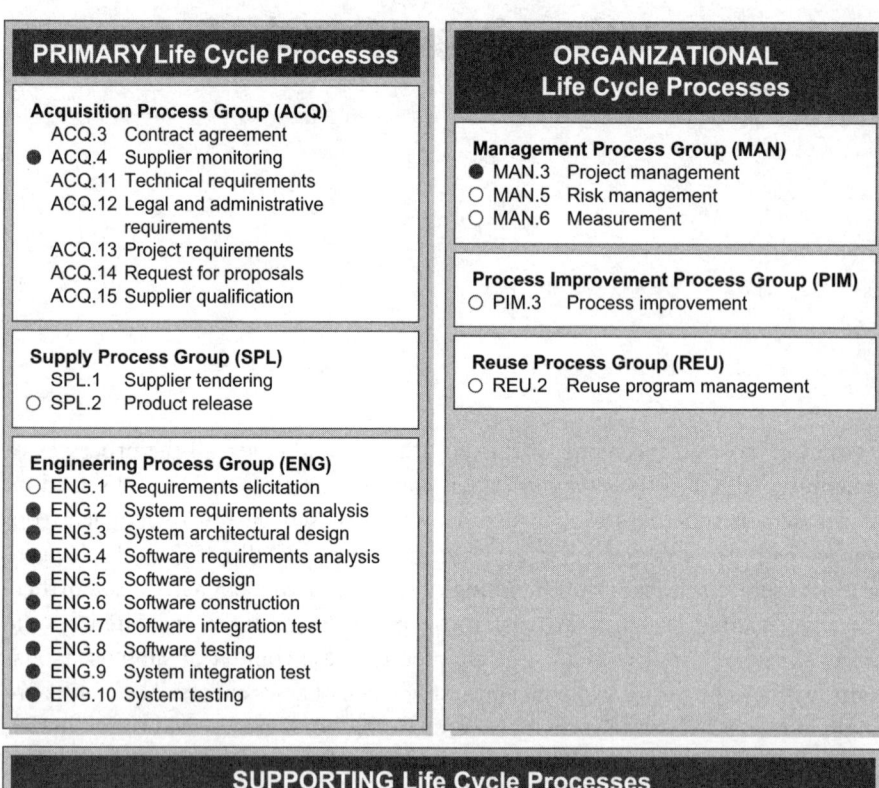

Figure 2–1   *Processes covered in this book (marked with bullets); filled bullets indicate processes required by the OEMs (so-called HIS scope). In case of software-only vendors, ENG.2, ENG.3, ENG.9, ENG.10 are not applicable. If no subcontractors are used, ACQ.4 is not applicable.*

The process purpose and the process outcomes are of particular importance because they are addressed by the Process Attribute PA 1.1 and the Generic Practice GP 1.1.1 (see chapter 3); as a result, they serve as the basis for capability assessments.

2 Interpretations Regarding the Process Dimension

Each of the marked processes in this chapter is given its own section which is divided into the following subsets:

- **Purpose**: The »process purpose« as specified in Automotive SPICE.
- **Characteristics particular to the automotive industry**: This section discusses the particular nature of automotive software and system development in relation to the process.
- **Base Practices**: A step by step, detailed discussion of the base practices defined in Automotive SPICE. In Automotive SPICE, notes are of particular relevance. In most cases they constitute concrete examples or interpretational guidance added by OEM members of the Automotive SIG and are frequently checked in OEM assessments.
- **Designated Work Products**: Automotive SPICE defines a large number of work products. We have added explanations and, to some extent, examples to those work products that are essential in practice, always referencing the work product identifier and work product name. In some cases—where Automotive SPICE does not provide a relevant work product—we have suggested such work products ourselves on the basis of our own experience.
- **Characteristics of Level 2**: Since the generic practices (see chapter 3) are not defined in a process-specific fashion, we have provided process-specific interpretations to help readers gain a practical process understanding. Generally, we have confined ourselves to Level 2 since the differences at Level 3 are not so significant. Using the subheadings »Performance Management« and »Work Product Management«, we refer to the Process Attributes PA 2.1 and PA 2.2.

In addition, we use irregularly occuring design elements:

- **Excursus**: We sometimes provide additional or cross-process comments in the form of excurses.
- **Notes to Assessors**: These notes not only give practical hints for use in assessments (e.g., in the form of typical assessment questions) but also point out particular problems, frequent weaknesses, and difficult rating situations.
- **Experience Reports**: Here we describe typical real-life problems or situations.

The term »customer« will be used quite frequently in this book and requires some additional explanation: Automotive SPICE uses »customer« to describe the relationship between two business partners, one being the provider or supplier and the other the recipient of a particular (development) product. In the automotive industry, »customer« is used to denote several different relationships. Using the term for internal relationships, vehicle manufacturers may use »customer« to refer to a particular product line organization or a different department. With regard to external relationships, »customer« is commonly used to denote the end user, the car buyer. On the suppliers' side, the customer is usually the vehicle manufacturer (OEM) or, in case of supplier chains, another supplier[2].

To help fully understand the engineering processes, some important key concepts are described in Annex D of the Automotive SPICE PAM and illustrated in figure 2–2[3]. These concepts are an essential basis for the understanding of the interaction of the engineering processes and corresponding Automotive SPICE work products. The central process is requirements elicitation (ENG.1), where customer requirements, system requirements (ENG.2) and software requirements (ENG.4) are collected. The system architectural design process (ENG.3) divides system requirements into mechanical, hardware and software requirements[4]. Software requirements specify a system's software with regard to its functional requirements[5]. The software design process (ENG.5) splits software up into software components and then into software units. During software integration (ENG.7), individual software units are integrated into software components which in turn are integrated to make up the overall integrated software. System integration (ENG.9) subsequently integrates the individual mechanical, hardware, and software items into the overall integrated system. Verification, e.g., test-based verification, always uses the verification criteria of the associated requirements. Finally, acceptance tests within system testing (ENG.10) validate the (software) system.

---

2. For example, if in a supplier chain a tier 2 supplier works for a tier 1 supplier and both are commissioned by the OEM.
3. In the original Automotive SPICE chart some of the words are illegible; this was compensated for in figure 2–2.
4. According to the IEEE definition, hardware denotes the electronic components for processing, storage, and transfer of programs or data. Mechanical components, for example, are housings, engines, levers, and fixtures.
5. For nonfunctional requirements see REU.2 and SUP.1.

2 Interpretations Regarding the Process Dimension

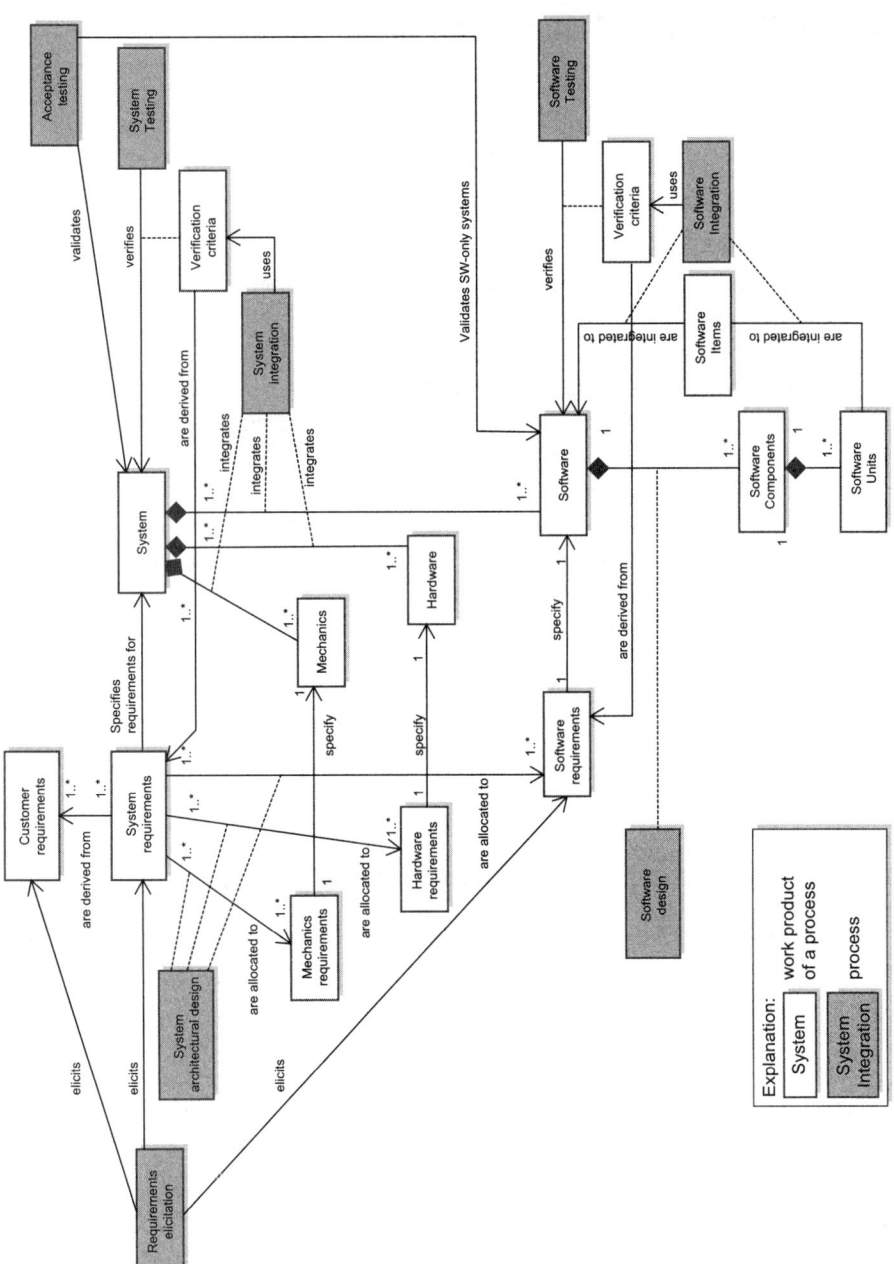

Figure 2–2   Key concepts of the engineering processes in Automotive SPICE

2.1 ACQ.4 Supplier Monitoring

2.1.1 Purpose

The purpose of the supplier monitoring process is to monitor the performance of the supplier against agreed requirements.

Besides discussing supplier monitoring, this process also deals with cooperation and communication with the supplier. The basis for cooperation is that a supplier selection has taken place and that a contractual agreement exists between the customer and supplier.

Methods from the MAN processes and SUP processes can be applied for supplier monitoring, such as project management, risk management, measurement and change management, except that they are applied here by the customer to monitor and control the supplier.

If development is contracted out to the supplier, the process interfaces must be coordinated. Besides the engineering processes, support processes in particular, such as configuration management, change management and quality assurance, should be coordinated.

For further information on supplier management see:
- [SA-CMM 1999] Software Acquisition Capability Maturity Model
- [DoD 1998] Software Acquisition Best Practices Initiative
- [PPSM 1998] Process Professional Supplier Management part 1-7

2.1.2 Characteristics Particular to the Automotive Industry

During the development of ECUs, vehicle manufacturers and suppliers frequently work in close cooperation. In some cases, development is done almost exclusively by the supplier, whereas other vehicle manufacturers may purchase already developed components requiring only small adaptations.

Sophisticated electronic vehicle components are often developed by several collaborating suppliers. Such networked developments frequently result in business partnerships. Typically, one of the suppliers is commissioned to act as the system supplier whose tasks include controlling the other suppliers.

Quite often this results in a whole hierarchy of supplier relationships, i.e., where one supplier (»tier one«) acquires further system components from his own subcontractors (»tier two«) but remains in control of the cooperation. Often a customer stipulates that the supplier subcontracts particular tier-two suppliers. In vehicle development, the selection and control of suppliers is generally of particular significance.

2.1 ACQ.4 Supplier Monitoring

> **Note to Assessors**
>
> This process is only assessed in an OEM supplier assessment if the supplier (tier-one) has subcontractors (tier-two). It assesses how the supplier monitors and controls its own suppliers. The process is applied if development is subcontracted or if product components are purchased that need to be adapted to a project's requirements. If only resources are purchased[a] for development activities (often called »body leasing«) or if standard products are purchased that do not require any adaptations (»COTS products«[b]), assessment of the process does not make sense. This is because ACQ.4 requires a strong interconnection of the development processes, which in these cases does not apply.

a. In this case leased staff work in accordance with customer processes; if required, they are questioned in an assessment in the same way as internal staff.
b. COTS Commercial off-the-shelf.

2.1.3 Base Practices

BP1: Agree on joint processes and joint interfaces. Establish an agreement on joint processes and joint interfaces, responsibilities, type and frequency of joint activities, communications, meetings, status reports and reviews. Agree on processes and interfaces at least for change management, problem management, quality assurance and customer acceptance.

NOTE: *The term customer in this process refers to the assessed party. The term supplier refers to the supplier of the assessed party.*[6]

During project execution the processes and interfaces between the customer[7] and supplier must be coordinated and documented. This comprises the following:

- Agreements regarding regular meetings and reviews (see BP2 to 4)
- Planning and control of communication and interfaces, e.g., by means of a communication plan (see figure 3–5 and chapter 3, deliberations on GP 2.1.6)
- Rules concerning the exchange of work products and information, e.g., which work products are to be made available, data exchange formats, method of transmission (e.g., e-mail encryption or communication via a common server with mutual access to joint project data and documents)
- Coordination of roles and responsibilities
- Planning and coordination of processes and work flows for common activities. The base practice requires this at least for change management and problem resolution management, quality assurance, and acceptance. It is also

---

6. For instance, if a tier-one supplier is assessed, the primary concern is how this supplier monitors and controls its own (tier-two) suppliers.
7. The customer can be the ordering party or not , e.g., if the OEM is formally the ordering party to the tier 2 supplier and obligates the tier 1 supplier to control the tier 2 supplier.

advisable to have rules for joint tests—at least regarding a jointly coordinated test strategy—as well as rules for configuration management, including release planning of customer supplied products and for consistent traceability across organization boundaries (see also section 2.24).

- Definition of escalation paths to handle problems; this is of particular importance if more than one supplier is involved.[8] A steering committee comprised of representatives of all involved parties may serve as the highest decision-making body.
- Tracking of open issues, for instance, using a common open issues list (OIL, see figure 2–3). In case of several suppliers the same problems may arise as explained in the footnote to the previous bullet.
- Rules regarding project status reports issued by suppliers to the customer and the exchange of project plans
- Cooperation between the customer and supplier[9] regarding quality assurance

It is good practice to set up a joint project team between the customer and supplier. This way dependable (major) roles (e.g., engineers in charge of ECU function development) can be assigned on both sides who can cooperate right from project start.

BP2: Exchange all relevant information. *Establish and maintain communications between customer and supplier for all agreed information, processes and interfaces.*

As agreed in BP1, information is regularly exchanged. Agreed communication must be retained throughout the project's duration.

BP3: Review technical development with the supplier. *Review development with the supplier on the agreed regular basis, covering technical aspects, problems and risks.*

Permanent communication in the project must be kept up, especially where technical aspects are concerned, to ensure that the supplier understands the technical requirements and that all work is done properly and according to plan. This is primarily done by means of regular project meetings focusing on technical issues, problems, and risks. In most cases there are a lot of technical questions and prob-

---

8. One frequent problem is that the responsibility for detailed analysis and correction of defects is ping-ponged between several partners, one passing the buck to the other. The customer must pay particular attention to this. It makes sense to establish an across-the-board defect management process that all suppliers have to apply and to which they have to jointly assign one person with overall responsibility.
9. In connection with this one must take care that the quality assurance and quality goals of the subsuppliers are consistent with those of the customer, i.e., the customer's qualitative requirements can only be extended but not be restricted by a subsupplier's local guidelines.

lems to be clarified during the course of a development project and all of them, according to BP5, must be tracked to completion.

Moreover, technical work products and documents should be evaluated in joint technical reviews, especially with regard to the following types:

- Customer requirements specification / System requirements specification
- System architecture (in the case of system suppliers)
- Software architecture
- Interface specifications
- Test plans
- »Test readiness«
- User documentation and other development documents that have been agreed upon
- Acceptance test (regarding customer acceptance)

In our opinion it is beneficial to conduct technical reviews with appropriate review methods (e.g., inspections) as early as possible[10] during the development processes (see also SUP.4).

BP4: Review progress of the supplier. *Review progress of the supplier regarding schedule, quality and cost on the agreed regular basis, also tracking problems to successful completion and performing risk mitigation activities.*

Supplier progress is regularly checked.[11] Project progress is checked against the plan and significant problems (including quality problems), risks, and schedules (in particular deadline shifts and forecasts) are discussed. If work is billable on a time and material basis, accrued costs are also checked against the cost plan. In addition, cost projections are evaluated.

Automotive SPICE requires furthermore that mitigation activities are initiated for identified risks and that identified problems and measures are tracked to completion (see BP5). Risk management (see MAN.5) in the project must therefore be extended to also cover risks that arise from a customer's cooperation with the supplier.

This practice gains particular relevance if several suppliers are to be be integrated and coordinated in a customer project. In that case both the overall project status of all supplier activities and the activities of the customer must be combined and monitored as a whole. Open communication of the overall status (to all stakeholders) is advisable in this case so that everyone understands their contribution to the overall effort. This transparency has the added advantage

---

10. This means immediately after creation and prior to release of the review object.
11. It helps a project with subsuppliers if they themselves provide relevant information on a regular basis (push principle) rather than having to react to the customer requesting it (pull principle).

that it indirectly creates competition which promotes project progress (nobody wants to be last). In order to be able to objectively track a supplier's project progress we recommend progress tracking by means of metrics (see MAN.6 and figure 2–27).

Progress monitoring should also include monitoring of agreed improvement measures (BP6) in the running project.

Experience Report

Suppliers often present project progress in a very optimistic light. Customers should, therefore, critically question these estimations (e.g., in status reports or in the project plan). Moreover, initial deliveries should be agreed upon early to allow for better tracking of the current project status.

Frequently, problems are caused by poor customer cooperation; often the supplier has to wait a long time for an answer (e.g., to technical questions), it may take a long time for decisions to be made, or urgently required domain experts are unavailable or only partly available. Another problem with networked development is the fact that components provided by the customer are often not available on time or insufficient (e.g., prototypes for test purposes).

BP5: Track open items. *Record open items found, pass them to the supplier and track them to closure.*

In order to ensure that all identified open items are closed as planned, a common open issues list (OIL, see figure 2–3) should be used to allow systematic tracking of such points.[12] This list can be used to track open issues of both the customer and supplier.

No.	Category	Issue Description	Action/ Solution	Prio.	Respon-sibility	Due Date	Status	Entry Date	Entry by	Source	Classifi-cation
1	CAN	problem description, e.g., CAN signal not on ...	description of required action	high	Jackson	03.28.2008	in progress	02.15.2008	McCabe	e.g., risk management workshop	change request
2	HMI			medium			open				open issue
3	trial			low			closed				defect
4	–										
5											

Figure 2–3   *Example layout of an open issues list (OIL)*

---

12. For systematic problem tracking see also SUP.9 BP8.

2.1 ACQ.4 Supplier Monitoring

BP6: Act to correct deviations. Take action when agreed targets are not achieved, to correct deviations from the agreed project plans and to prevent reoccurrence of problems identified.

If, in the course of a project and during monitoring of its progress, deviations from the plan are noticed which indicate that agreed targets cannot be met, appropriate measures must be taken. Besides requiring measures to correct such deviations from the plan, Automotive SPICE also requires measures that remove or obviate their causes.

BP7: Agree on changes. Changes on agreed activities proposed by either party are negotiated and the results are documented in the agreement.

A defined and systematic change management process (see SUP.10) must be deployed between the customer and supplier for the entire lifecycle of a project. Besides the coordination of changes, documentation of the results is also required. It must be pointed out here that changes may not only involve technical questions but also defined processes, practices, agreements, contracts, etc.

Experience Report

In fixed-price projects, the customer-supplier relationship often suffers if the implementation of changes causes cost increases for the supplier. For this reason many customers shy away from issuing formal change requests; instead, requests for changes are passed on to the supplier informally or are concealed as error reports. In system development this situation is aggravated if new technologies are being applied and if the project outcome is the result of an evolutionary process. In this case requirements can only be defined insufficiently (i.e., incomplete or with insufficient detail) and can only be passed on bit by bit to suppliers during development. In such cases the change management process must be adapted accordingly. For instance, an agreement may be made that change requests are formally introduced via the change management process but that they only become cost effective under specific conditions.[13]

---

13. For example, changes may become cost-relevant if a change is of a functional nature and if the associated subsystem has already been provisionally accepted.

2.1.4 Designated Work Products

02-01 Commitment/Agreement

A valid and binding agreement of any kind. The agreement must be documented for it to become verifiable and binding to all parties (e.g., as part of agreed minutes and perhaps even confirmed by signature). ACQ.4 refers to agreements between customer(s) and supplier(s).

13-01 Acceptance Record

According to [Hindel et al. 2006], the acceptance record consists of the actual record itself and an attached list of deficiencies. Basic information on the list of deficiencies can be found in SUP.9, work product 15-12 (problem status report). The acceptance record should contain the following additional information:

- Project data (project name, order number, order date, customer, etc.)
- List of all products to be accepted as per release planning, in addition to the actual (sub)system, also all delivered documentation and accompanying material (recipient, delivery date, identifier including version information, etc.)
- Date, location, parties involved in the acceptance process (including allocation to either supplier or customer)
- Type of acceptance (for example, partial acceptance or final acceptance)
- Reference to other input and/or output documents associated with the acceptance
- Test or verification results
- Acceptance result (and, if required, next steps such as defect correction or renewed acceptance)
- Signatures of both the customer and supplier

13-09 Meeting Support Record

Meetings between customer and supplier should be documented using meeting minutes containing the following:

- Meeting purpose, agenda, location, date and time, and participants
- Decisions made and other relevant results
- Open issues (if required, this can be a separate list, see figure 2–3)
- If necessary, date of the next planned meeting
- Distribution list

13-14 Progress Status Record

See MAN.3, work product 15-06 (project status report).

13-16 Change Request

For explanation see SUP.10

13-17 Customer Request

A change request issued by the customer to the supplier. For explanation see SUP.10

2.1.5 Characteristics of Level 2

On Performance Management

At project start, communication between the customer and the supplier must be coordinated and planned in line with BP1 (coordination of process interfaces, coordination of roles and responsibilities, definition of escalation paths, etc.). During the course of the project all regular meetings, progress reviews, and acceptances (partial and final) are planned and their performance and compliance tracked.

On Work Product Management

All supplier work products that were agreed to according to BP1 as part of the communication process are to be reviewed and put under version and change managment control. In addition to the agreed intermediate and final deliveries, this also applies to agreed planning documents and progress reports of the supplier. If required, change requests are to be reviewed also. One may also want to agree on the establishment of a joint configuration management system, and that work products are versioned and changed accordingly.

2.2 SPL.2 Product Release

2.2.1 Purpose

The purpose of Product release process is to control the release of a product to the intended customer.

A product release is a consistent set of versioned objects with defined attributes and features designed to be delivered to an internal or external customer. In terms of configuration management a product release therefore constitutes a baseline (see SUP.8 BP5). The process contains the following:

- Planning and control of releases
- Determination of release criteria
- Preparation and implementation of intermediate and final releases
- Compilation of »build lists«
- Delivery method
- Release documentation and support

The process has interfaces to requirements management, project management, and configuration management. All requirements to be implemented are prioritized during requirements management (see, for instance, ENG.2 BP4). During project management these requirements are assigned to different releases and milestones using, for instance, a function list (see figure 2–19). Implementation is subsequently tracked.

2.2.2 Characteristics Particular to the Automotive Industry

The automotive industry works with varying prototypes called samples (A, B, C, D). Components under development are provided as prototypes with increasing functionality and then integrated in test vehicles. The aim is to integrate early and to test their behavior in the vehicle. The different sample stages are defined as follows:

- **A-samples:** These are functional prototypes with limited drivability and a low degree of maturity. The component does not yet fulfill all the requirements regarding, for instance, temperature and voltage range, dimensions, impact and vibration resistance, electromagnetic compatibility (EMC) and visual appearance. However, A-samples allow early testing of basic (software) functions under test conditions. A-samples are used to illustrate functionality.
- **B-samples:** These are functional, basic prototypes with full drivability and a high level of maturity. A B-sample looks like the later series component and has all the (software) functionality required at that stage of development. B-samples are, to a large extent, composed of pre-series hardware and allow first statements regarding electromagnetic compatibility and temperature range, although they may have been created using pilot tools. B-samples ensure sufficient operational safety for trials at the test bench and in the vehicle and contain an infrastructure for diagnosis. However, there may still be restrictions for their deployment in the car. B-samples are used to define construction issues, such as assembly and space requirements, and are used for function tests, function validation, and calibration either in the car, test bench, or test lab.
- **C-samples:** C-samples are manufactured under pre-series conditions using series production tools and are used, among other things, for endurance testing. Component construction and specifications fulfill the requirements of the

2.2 SPL.2 Product Release

series product in that all specifications regarding function, reliability, and interference resistance are complied with. Moreover, installation dimensions, space requirements, and contacts correspond to the series product. Furthermore, reproducibility during series production must be guaranteed. In C-samples, no technical constraints are permitted, enabling unrestricted use in the vehicle. The electronic components must be genuine series components. In theory, all software functions should already be available in the B-sample; in practice, however, some completed software functions of some ECUs will only be ready with the C-sample. C-samples are used in overall testing (endurance test, function validation, calibration) under pre-series conditions.

- **D-sample:** These are the final prototypes provided by the supplier for the design sample release. Design samples, sometimes called D-samples, are produced using series production tools under conditions similar to series production. The systems are fully operational and can be evaluated. All quality requirements are supposed to be consistently met. Released design samples can also be used for the first sample release and be fitted in pre-series/series production.

2.2.3 Base Practices

BP1: Define the functional content of releases. Establish a plan for releases that identify the functionality to be included in each release.

Release planning defines which particular features will be implemented in which product version. Based on this plan, the development process can be structured and tasks can be prioritized. Function lists (see figure 2–19) are often used to plan the functional content of each release. Intermediate releases may also be necessary (e.g., hardware status B-sample but including latest software functionality) in addition to the »officially« agreed A/B/C-samples, and these must be planned, too. Release planning changes quite frequently during the course of the project. This may be for the following reasons:

- The project is delayed but release dates must be met. As a result, the functionality of the release is reduced and an additional release with full functionality will be delivered later.
- Requirements change during the course of the project; they are re-prioritized or new requirements need to be considered; as a result, requirements are implemented in an earlier release than originally planned or released later.

Additional releases on top of those originally planned should be applied for and approved. A release request should at least contain the following items:

- Affected objects
- Reasons for the release
- The way in which the delivery will be performed

Release planning must be tracked and updated according to Process Attribute PA 2.1.

BP2: Define release products. The products associated with the release are defined.

NOTE: The release products may include programming tools where these are stated. In automotive terms a release may be associated with a sample e.g., A, B, C.

The objects and documents contained in a release (see also SUP.8 BP2) must be defined in addition to the definition of the functional content. These may, for instance, be documents associated with development, test records, QA reports, or a list of known defects. Furthermore, it must be defined if only a software release or a complete system (e.g., software running on specific hardware) will be delivered. Is the delivery of an agreed A/B/C-sample limited to a box of ECUs, or does it also comprise documentation and necessary tools, such as calibration tools or flash tools[14]?

BP3: Establish a product release classification and numbering scheme. A product release and classification is established based upon the intended purpose and expectations of the release.

NOTE: A release numbering implementation may include

- *the major release number*
- *the feature release number*
- *the defect repair number*
- *the alpha or beta release*
- *the iteration within the alpha or beta release.*

A classification scheme is established. Possible classifications include:

- Internal release
- Customer release for test purposes
- Official sample release
- Series release

Moreover, a numbering scheme is introduced so that releases can be provided with a unique identifier, e.g., T02A (trial phase, 2nd sample release A), thus making clear communication a lot easier.

---

14. Calibration tools are used to change software parameters; flash tools are used for flashing new software versions into the ECU

2.2 SPL.2 Product Release

BP4: Define the build activities and build environment. *A consistent build process is established and maintained.*

NOTE: *A specified and consistent build environment should be used by all parties.*

During release planning the compilation of each release build also needs to be defined. This is usually specified via configuration management (see SUP.8). If several suppliers are involved, the build process may be more complex. In this case a procedure describing how to compile the build must be agreed on. Is, for instance, the system supplier going to carry out the build? Are there several stages, etc.?

In addition to defining how to proceed, the build environment, like the compiler, target link version, Matlab version, etc., needs to be coherently defined and applied. These definitions or rules must be adapted during the project's lifetime whenever necessary.

BP5: Build the release from configured items. *The release is built from configured items to ensure integrity.*

NOTE: *Where relevant the software release should be programmed onto the correct hardware revision before release.*

The objects defined in BP2 are compiled according to the procedure defined in BP4. All objects are obtained without exception from the configuration management system.

BP6: The type, service level and duration of support for a release are communicated. *The type, service level and duration of support for a release is identified and communicated.*

Support is arranged for the delivered sample. Among the conceivable items are arrangements regarding support concerning changes, procedures in case of defects (e.g., new delivery in case of one or more priority 1 defects), immediate provision of development resources if necessary, provision of a resident engineer, support regarding software-only deliveries (e.g., navigation software, operating system) via hotline, etc.

BP7: Determine the delivery media type for the release. *The media type for product delivery is determined in accordance with the needs of the customer.*

NOTE: *The media type for delivery may be intermediate (placed on media such as floppy disk and delivered to customer), or direct (such as delivered in firmware as part of the package) or a mix of both. The release may be delivered electronically by placement on a server. The release may also need to be duplicated before delivery.*

Possible media devices can be:

- Magnetic tapes, CD, DVD, etc.
- Electronic distribution, e.g., via the internet or e-mail
- Mail, parcel services, or personal delivery

BP8: Identify the packaging for the release media. *The packaging for different types of media is identified.*

NOTE: The packaging for certain types of media may need physical or electronic protection for instance floppy disk mailers or specific encryption techniques.

Besides determining the delivery medium, arrangements regarding packaging of the delivery may also be necessary:

- Sensitive hardware or mechanical parts must be specially packaged to avoid damage during shipping.
- How many components are included in each delivery?
- Are there safety requirements? For instance, how are connections to be secured or cables to be fastened?
- Are protective measures such as encryption techniques to be applied for electronic deliveries?

BP9: Define and produce the product release documentation / release notes. *Ensure that all documentation to support the release is produced, reviewed, approved and available.*

In addition to the release description, accompanying release documentation may comprise the following:

- Was functionality implemented according to plan?
- Which known defects still exist in the release; does it include workarounds?
- What is the memory consumption?
- Was comprehensive (acceptance) testing performed (e.g., including tests regarding temperature and voltage range, impact and vibration resistance, EMC, etc.)?
- What is the release status (full release, conditional release, etc.)?

BP10: Ensure product release approval before delivery. *Criteria for the product release are satisfied before release takes place.*

During approval, fulfillment of the defined release criteria must be checked. In practice this is normally a formal act by the project management or management, since it is usually known beforehand whether a release can be made or not.

2.2 SPL.2 Product Release

BP11: Ensure consistency. *Ensure consistency between software release number, paper label and EPROM-Label (where relevant).*

Prior to shipment a check needs to be carried out to see if in fact the release contains all objects identified in BP2 (»Do we get what's written on the package?«) and that labels and the software release number are consistent.

Quite problematic in practice is the possibility that the software release number may not correspond to the actually shipped software. The same applies to modification labels on the sample. Both issues are addressed in SUP.8 BP10.

BP12: Provide a release note. *A release is supported by information detailing key characteristics of the release.*

NOTE: *The release note may include an introduction, the environmental requirements, installation procedures, product invocation, new feature identification and a list of defect resolutions, known defects and workarounds.*

For each release, release documentation (e.g., release note, development report, etc.) is issued. Regarding the contents of the accompanying release documentation see BP9.

BP13: Deliver the release to the intended customer. *The product is delivered to the intended customer with positive confirmation of receipt.*

NOTE: *Confirmation of receipt may be achieved by hand, electronically, by post, by telephone or through a distribution service provider.*

NOTE: *These practices are typically supported by the SUP.8 Configuration Management process.*

NOTE: *Refer to ISO/IEC 9127: 1988 'User Documentation and Cover Information for Consumer Software Packages' for guidance on packaging aspects of software product supply.*

The product and associated documentation are delivered and the customer confirms their receipt. In some cases the delivery may also include the installation of the product in a defined environment.

2.2.4 Designated Work Products

08-16 Release Plan

Typically, the release plan contains the following:
- The planned schedule of all releases (overview)
- Rules covering the handling of the releases, e.g., classification, numbering scheme, build activities and build environment, etc.

Additionally, rules are needed regarding the following:

- The functional content (see function list, figure 2–19)
- Associated objects (software version, hardware version, documentation, etc.) and releases
- A mapping onto the customer requirements implemented in each of the releases
- Additional miscellaneous regulations regarding support, packaging, type and medium of the delivery

2.2.5 Characteristics of Level 2

On Performance Management

At Level 2, release planning and release tracking are managed much more stringently and with more detail (Process Attribute PA 2.1).

On Work Product Management

The requirements of Process Attribute PA 2.2 particularly apply for the release itself and all associated release objects. Checking of the release is done within the context of acceptance testing. Furthermore, the release plan should be under configuration control.

2.3 ENG.1 Requirements Elicitation

2.3.1 Purpose

The purpose of the Requirements elicitation process is to gather, process, and track evolving customer[15] needs and requirements throughout the life of the product and/or service so as to establish a requirements baseline that serves as the basis for defining the needed work products.

If requirements are not elicited with sufficient care, the whole project is left without a solid basis and later problems are bound to occur. Documenting the requirements in writing often causes difficulties in practice or is deficient. Common reasons for this are:

- The development project is under enormous time pressure right from the start and as a result, requirements are poorly defined.
- The customer cannot or is not willing to commit to all the details early in the project.

---

15. »Customer« may be the OEM or another supplier (at a level above), independent of the fact of who is formally the ordering party.

- A highly innovative and complex system is being developed for the first time and initially the requirements are difficult to define.

Generally, elicitation of requirements during the quotation phase suffers from considerable time pressure. Especially in a customer-supplier relationship, the analysis of requirements cannot be performed as profoundly as is necessary because the experts are forced to confine themselves to the very basics under the pressure of the quotation's looming due-date. Under such circumstances detailed analysis can only be conducted after an order has been placed, which in turn leads quite often to differences of opinion between the partners about the quotation's content. The following scenarios have proven to be problematic for the further course of the project:

- The customer is not involved in the specification of the system function and the specification effort is left entirely to the supplier. However, the customer issues a variety of change requests and new requirements later in the project.
- The supplier fails to have requirements it specified confirmed by the customer, or the customer refuses confirmation.

In some cases a »labor-saving practice« has become established whereby a system is ordered »identical to that of the competitor« or »like the one in the predecessor project, with only the following changes ...«. At this point in the project the seed for later problems is already sown.

Using existing and already developed platform solutions, or reusing applications designed for other customers, can also be problematic. Such constellations may appear quite enticing when placing an order; however, due to necessary changes and adaptations they can later quickly drive a project to the brink of failure and later raise the question of the economic competitiveness of reuse. In practice it has happened more than once that companies offering platform solutions had to scrap them during the project and go right back to the drawing board. The reasons for this may be that changes can become too extensive in relation to a project's size or that the planned solution can no longer be implemented due to architecture problems. Success in these cases does not only depend on the identification of the requirements but on other factors, too, such as change management and its connection with the engineering processes, the software architecture, encapsulation of individual software components, etc. Consequently, customers will get either a »patchwork solution« of added changes or a consistent product that corresponds with their requirements.

In complex systems we often find a large number of requirements defined and spread over different documents. Consistency of the content of these requirements must be guaranteed throughout the project's lifetime despite any changes made. Frequently, changes are collected over a period of time and then included in the requirements documentation in one shot, a step that is often used to create so-called requirements baselines.

2.3.2 Characteristics Particular to the Automotive Industry

In the majority of cases requirements do not result from explicitly expressed customer needs alone. In complex products, additional requirements, standards, or other regulations such as diagnosis regulations and specifications related to ECU flashability must also be taken into account. The number of documents that need to be considered may very quickly become several hundred. In some cases documents are out-of-date and quite often they are riddled with contradictory information. This leads to a considerable development effort, since all these documents not only need to be examined, assigned, and prioritized, but all inconsistencies, errors, and technically infeasible requirements contained in them must be identified and sorted out. Customers are rarely aware of the complexity and effort that these activities entail, but in giving precise and sound specifications they can greatly contribute to the prevention of later problems.

2.3.3 Base Practices

BP1: Obtain customer requirements and requests. Obtain and define customer requirements and requests through direct solicitation of customer input and through review of customer business proposals, target operating and hardware environment, and other documents bearing on customer requirements.

NOTE: *The information needed to keep traceability for each customer requirement has to be gathered and documented.*

Most commonly, the following cases will be relevant for requirements elicitation:

- A product is being developed to order. The requirements come–at least in part–directly from the customer. Applicable methods are, for instance, interviews, workshops, and the analysis of customer documents.
- A product is being developed directly for the market. In this case, requirements are specified by a company's own product marketing or product management, and the mechanisms and techniques applied differ from those in the first case, for instance, in the way market analyses, end customer surveys, or analysis of competing products are performed.
- A hybrid form of the first two cases is used to develop a specific product for a customer. However, it is supposed to be based on an existing product platform. Alternatively, the product is intended to serve as the basis for further products (therefore being the basis for a product platform).

In any case it is important to actively approach requirements providers and elicit their requirements. When identified requirements are documented, traceability (see ENG.2) must be ensured by indicating the source of the customer requirements (for instance, the customer concepts mentioned in the base practice, target operating environment and hardware environment, together with other docu-

2.3 ENG.1 Requirements Elicitation

ments which may have an impact on the customer requirements). In case of the hybrid form, customer-specific and platform-specific requirements need to be distinguished. Arising conflicts should be identified and eliminated early. System and software requirements are derived from the customer requirements at the subsequent levels.

> **Note to Assessors**
>
> One crucial (and rating-relevant) point regarding BP1(and BP2) is the supplier's active endeavor to elicit requirements and put them into the context of application. In practice, however, problems frequently arise from the involvement of the customer. The supplier should not be down-rated in an assessment if a customer is unable to convey his requirements or to actively participate in articulating them. Findings of this kind should, however, be mentioned in the report.

BP2: Understand customer expectations. Ensure that both supplier and customer understand each requirement in the same way.

NOTE: *Review with customers their requirements and requests to better understand their needs and expectations. Refer to the process SUP.4 Joint review.*

Joint reviews involving customer and supplier are an effective way to achieve a common understanding, as mentioned in the note. In iterative development models they are often done on the basis of incrementally enhanced prototypes and may be continued during a large part of the project's life.

Unclear or ambiguous requirements are typically identified by the supplier during the review of the requirements document in the domain departments and conveyed to the customer. Open issues can be clarified in a subsequent joint review. Such requirement reviews are not only conducted at project start but must be repeated whenever there are changes or enhancements to functions to ensure that the supplier really understands what the customer wants. At the same time the supplier must also check the feasibility and suitability (e.g., regarding comprehensibility and consistency). Review results should be documented and approved by both parties in order to avoid misunderstandings and conflicts.

Though not required by the base practice, implicit requirements (see BP4) should be treated in exactly the same way as explicit customer requirements, i.e., reviewed, agreed, approved, etc.

BP3: Agree on requirements. Obtain an explicit agreement from all relevant parties to work to these requirements.

The agreement of everyone involved in the development, i.e., on the supplier's and customer's side, requires that each team has assessed the requirements and considers them feasible. This way problems concerning technical feasibility and associated risks are considerably reduced and a common development objective is established. In practice, obtaining a binding agreement is usually difficult and is normally effected by management after expert evaluation. Particular emphasis is placed on the completeness of the information retrieved from the involved stakeholders. For this reason we need to identify and document at project start who the relevant stakeholders are and when they made which decision.

Often the commitment of the supplier's technical departments to implement the requirements of the customer is already obtained during the quotation phase (i.e., before the order is placed) by the quotation team or the (future) project management. As a result the supplier has gained an understanding of the customer's requirements and their feasibility. Based on a first assessment of the feasibility and a rough effort and cost estimation, a first offer is typically drawn up, the content and scope of which will be further refined and adjusted jointly with the customer in subsequent negotiations. If a detailed requirements analysis is not performed until after the order has been placed the commitment must be renewed based on the detailed understanding of the requirements.

This process, which is performed in several iterations at the supplier's, is not explicitly mapped in the Base Practices of Automotive SPICE. Likewise, Automotive SPICE does not explain how to deal with requirements that cannot be implemented. It is advisable that requirements that the supplier cannot agree to or which he cannot fulfill are identified and communicated to the customer in writing before the order is placed (»deviation list«). If this is not done the requirements become an integral part of the contract, which can lead to problems later on.

BP4: Establish customer requirements baseline. *Formalize the customer's requirements and establish as a baseline for project use and monitoring against customer needs. The supplier should determine the requirements not stated by the customer but necessary for specified and intended use and include them in the baseline.*

The baseline (see also SUP.8) constitutes both a baseline for the project and a basis for the management of changes. Along with the customer's requirements there are usually additional requirements, for instance, environmental regulations or obligations, bills and other legal requirements, standards or applicable regulations for different markets or regions that need to be taken into account. The supplier must concern himself with the full identification of all relevant requirements, whether explicitly coming from the customer or not.

Once the supplier has identified (BP1 and BP2) and agreed upon (BP3) all relevant requirements, the agreed status is retained to form the requirements baseline. This can be done either in the form of a requirements specification or as a baseline across all requirements documents in a configuration management system. Based on this initial version each change or enhancement of the customer's requirements is identified, evaluated, and tracked later in the project using a

2.3 ENG.1 Requirements Elicitation

change management procedure (see BP5). This way additional baselines are successively created.

Practice has shown that projects without requirements baselines are problematic because the supplier cannot identify if a requirement has already been considered and because changes are often not identified as such in the project.

The requirements specification should satisfy formal requirements as, for instance, expressed in IEEE 830-1998 »Recommended Practice for software requirements specifications«. According to the standard, requirements should be correct, unambiguous, complete, consistent, ranked for importance and/or stability, verifiable, modifiable, and traceable [IEEE 830].

The requirements baselines are used for the creation of additional documents during the course of the project (e.g., for the requirements specification), with more details being added over time (see ENG.2). As such, they are of essential importance for the project. One of the reasons why the first requirements baseline is so important in practice is because it constitutes the basis for the project plan. In a customer-supplier relationship the first baseline is often developed during the quotation phase and becomes an integral part of the contract.

BP5: Manage customer requirements changes. *Manage all changes made to the customer requirements against the customer requirements baseline to ensure enhancements resulting from changing technology and customer needs are identified and that those who are affected by the changes are able to assess the impact and risks and initiate appropriate change control and mitigation actions.*

NOTE: *Requirements change may arise from different sources as for instance changing technology and customer needs, legal constraints.*

A well-functioning requirements change management system is essential for the project. With this in mind it is necessary that change requests follow a defined work-flow (see also SUP.10). An example of a typical work flow would be as follows:

- Change requests are collected and passed on to an individual or group authorized to make a decision (often the project manager himself or together with other members of staff, perhaps even together with the customer). What must be avoided is that individual developers accept unauthorized changes »unbureaucratically«, so to speak, and then implement them, producing potentially undesired side effects and unplanned extra costs.
- Each change request is evaluated with respect to its impacts and feasibility.
- If in a customer-supplier relationship the first requirements baseline is an integral part of the contract, the supplier must be able to separate effort already commissioned from that which is additional. Each change or enhancement requested later by the customer is estimated and a decision made as to whether the change is chargeable or not, i.e., whether or not a supplemental offer must be issued.

- The change request is subject to risk analysis and, if needed, measures for risk reduction are planned (see also MAN.5).
- If the decision is positive the change effort is estimated, considered in the project plan, and internally commissioned. The customer is then notified. The change request is tracked all the way up to completion using the project's tracking mechanisms (see also MAN.3).
- A negative decision will be justified to the customer.

What needs to be taken into account in this respect is not only requirements of the customer or end customer but also all changes necessitated, for instance, by technology changes or changed legal requirements. The supplier must therefore observe the market continuously to obtain knowledge about these changes of the state of the art. Subsequent implementation of the relevant changes must be tracked during the course of the project to ensure that this is done as agreed and that all changes will be available for the planned date or release.

BP6: Establish customer-supplier query communication mechanism. *Provide a means by which the customer can be aware of the status and disposition of their requirements changes and the supplier can have the ability to communicate necessary information, including data, in a customer-specified language and format.*

NOTE: *This may include joint meetings with the customer or formal communication to review the status for their requirements and requests; Refer to the process SUP.4 Joint review.*

NOTE: *The formats of the information communicated by the supplier may include computer-aided design data and electronic data exchange.*

The customer gets informed about the status (e.g., has a decision been made concerning the change request?) and the disposition (e.g., in which version is the change going to be implemented?) of his changed/enhanced requirements. A defined communication mechanism is in place and used for status reporting. The necessary technical prerequisites (data connections, tools, format specifications, etc.) are provided. This is to ensure that the customer understands the reports and that he can process them further should the need arise.

If no specific tool is used the customer can, for instance, maintain a list of the change requests and ask the supplier to periodically add comments and complete the list. Moreover, it is common practice to have joint meetings with all stakeholders to discuss the status of requirements and change requests.

It is advisable that a proactive feedback mechanism to the customer is put in place to give him the necessary confidence that his requirement changes are treated in a competent manner within the agreed timeframe. This includes always keeping the customer informed about the degree of completion regarding his requirement changes.

2.3 ENG.1 Requirements Elicitation

The end customer is informed on an as-needed basis, e.g., through press releases on recalls and corrective actions, or, for instance, via commercials for new products in which the desired changes have been implemented.

2.3.4 Designated Work Products

13-21 Change Control Record
see SUP.10

17-03 Customer Requirements

The requirements of the customer define the features of a system or product. Depending on the project configuration they are either provided by the customer (in the form of a »customer requirements specification«) or developed within the project.

The following quality attributes characterize a requirement specification:

- Correctness: Regarding compliance to higher level requirements
- Unambiguity: Regarding language
- Completeness: Regarding functionality, performance, interfaces, etc.
- Verifiability: Requirements are clearly stated and measurable
- Consistency: Terms are used in a consistent manner (no synonyms), and requirements are not contradictory
- Modifiability: Content matter is structured, systematically organized (e.g., table of contents, index, cross-references), and consistent
- Traceability: Individual requirements are separated and forward traceable to subsequent documents and backwards traceable to sources (see section 2.24)

On the structure of requirements documents see figure 2–4.

```
1. Introduction                              3.x Functional Requirements (Function x)
   1.1  Purpose                                 3.x.1  Introduction
   1.2  Scope and Constraints                   3.x.2  Inputs
   1.3  Definitions, Abbreviations,             3.x.3  Flow Control
        List of Abbreviations                   3.x.4  Outputs
   1.4  References                              3.x.5  External Interface
   1.5  Overview                                       3.x.5.1  User interfaces
2. General description                                 3.x.5.2  Hardware Interfaces
   2.1  Product Overview                               3.x.5.3  Software Interfaces
   2.2  Product Functions                              3.x.5.4  Communication Interfaces
   2.3  User Characteristics                    3.x.6  Performance Requirements
   2.4  Restrictions                            3.x.7  Design Constraints
   2.5  Preconditions and Dependencies          3.x.8  Characteristics
3. Specific requirements                                3.x.8.1  Safety
   3.1  Functional Requirements (Funct. 1)              3.x.8.2  Maintainability
   3.2  Functional Requirements (Funct. 2)      3.x.9  Other Requirements
   3.n  Functional Requirements (Funct. n)             Data basis, Expiry Date ...
```

Figure 2–4 *Example of a requirements document structure according to [IEEE 830]*

2.3.5 Characteristics of Level 2

On Performance Management

The content matter listed in Process Attribute PA 2.1 is usually implemented using the project's project management methods and partly documented in the project plan. Planning of activities within change management is rather confined to scheduling meetings and to the procedures mentioned in BP5.

On Work Product Management

The requirements of Process Attribute 2.2 particularly apply to the requirements document and change records.

2.4 ENG.2 System Requirements Analysis

2.4.1 Purpose

The purpose of the system requirements analysis process is to transform the defined customer requirements into a set of desired system technical requirements that will guide the design of the system.

The system requirements analysis is one of the most important processes since it constitutes a vital basis for the entire development work. System requirements describe the requirements for the overall system, consisting of different hardware, mechanics, and software components, and for the interaction of these components. A poor system requirements analysis is one of the biggest failure factors in development projects.

ENG.2 takes up the customer requirements identified in ENG.1 (= customer requirements specification level) and translates them into technically more detailed requirements (= system requirements specification level). In addition to the customer requirements, the requirements of all the other groups and individuals involved in the development are taken into account. Particularly critical for a successful implementation of the system requirements analysis is the interaction of the different development areas. such as hardware, mechanics, software development, and test departments.

2.4.2 Characteristics Particular to the Automotive Industry

It is often difficult to separate requirements for individual system components from the customer requirements documents, where the system's desired functionality is usually in focus from the end customer's point of view. In system requirements analysis, focus is now on the technical view of the supplier's development team. In practice (though not required in Automotive SPICE), requirements are

2.4 ENG.2 System Requirements Analysis

often separated for individual components. This may, for instance, be done by a corresponding structuring of the system requirements documents or by separating them into different documents.

Excursus: System

Most assessments begin with the question: »How do we define the term system?« If this question is not unambiguously answered it will lead to repeated confusion and negative effects later in the assessment. AUTOSIG (the Automotive Special Interest Group) themselves needed two days for the discussion and unambiguous definition of this term. We need a precise definition because subsequent steps (e.g., ENG.4 Software requirement analysis) build on this system definition. For this reason, at least a clear differentiation between hardware and software should have been made.

[IEEE 610] defines a system as a »collection of components organized to accomplish a specific function or set of functions«. [DIN EN 61508] defines the system as a » set of elements that interact according to a design, where an element of a system can be another system, called a subsystem, which may be a controlling system or a controlled system and may include hardware, software, and human interaction«. Systems in the automotive industry can be composed of different mechanics, hardware, and software components. Besides others, the following definitions of system are possible:

- The vehicle with all its individual components, which themselves can be subdivided into smaller subsystems.
- Assembly part or module consisting of mechanics (e.g., steering unit), ECU hardware, and software. This definition corresponds to the one provided by AUTOSIG.
- ECU consisting of mechanics (housing), hardware (printed circuit board, electronic components), and software. This definition only refers to the ECU and thus constitutes a subcategory of the definition above.
- Vehicle function distributed over several components, modules, or ECUs. In this case system consideration may become as complex as you like in the course of an assessment.
- Software function distributed over several control units.

2.4.3 Base Practices

BP1: Identify system requirements. *Use the customer requirements as the basis for identifying the required functions and capabilities of the system and document the system requirements in a system requirements specification.*

NOTE: *System requirements include: functions and capabilities of the system; business, organizational and user requirements; safety, security, human-factors, engineering (ergonomics), interface, operations, and maintenance requirements; design constraints and qualification requirements (ISO/IEC 12207).*

NOTE: *For system requirements specifications, the IEEE-Standard 1233-1998, Guide for Developing System Requirement Specifications might be used.*

The system requirements are derived based on the customer requirements in ENG.1. Often function[16] or feature lists are created that form the basis for the later system requirement specification. Initially there is often no specification of how system requirements are to be implemented. According to the first note above, system requirements analysis not only comprises the function requirements but is extended by additional factors important to the customer and the end customer. System requirements analysis is easier if the requirements gathering was tool supported and if analyses can be done accordingly. Generally, there is a noticeable trend in this direction. The standard tool currently used by German vehicle manufacturers is DOORS.

The result of a system requirements analysis is a baseline of the system requirements (e.g., as a version of a system requirement specification) that is updated and extended during the further course of the project. A »baseline«[17] constitutes an agreed working basis for the subsequent development phases and is drawn at defined milestones or dates in the project. System requirements baselines should be under configuration control or at least be versioned. In relation to the system, the following requirements should be considered:

- Functional requirements and features
- Interfaces, system performance, timing behavior, requirements for operational use, and design restrictions/constraints
- Nonfunctional requirements; these bear no direct relation to the technical function of system elements. Among these, for instance, are quality requirements, modifiabiliy, testability, maintainability and reliability, documentation, commenting, and verification (e.g., coming from coding guidelines), commercial constraints (such as, for instance, a customer's business and organizational requirements), business-management aspects, market requirements, time-to-market, reuse, maintenance, and ongoing product support.

---

16. See also the function list in MAN.3, figure 2–19.
17. See also SUP.8.

2.4 ENG.2 System Requirements Analysis

- Technical constraints
- Standards (ISO, DIN, etc.) to be taken into account, e.g., regarding functional safety [DIN EN 61508], customer standards (e.g., technical guidelines)
- Data security
- Ergonomics

Before requirements become contractual their feasibility may have to be analyzed (to this end, see BP2).

BP2: Analyze system requirements. *Analyze the identified system requirements in terms of technical feasibility, risks and testability.*

NOTE: *The results of the analysis may be used for categorization of the requirements (see also ENG.2.BP.4).*

The system requirements identified in BP1 are analyzed for feasibility, risks, and testability. System requirements that can be traced back to customer specifications (typically in the form of a customer requirements specification) are in the center of the technical feasibility analysis. The supplier's technical departments involved in the development evaluate the feasibility of the project, i.e., they determine to what extent the system requirements are feasible from a technical point of view. Among others, considerations regarding feasibility include the following questions:

- Are the system requirements and other assumptions realistic?
- Are there alternative solutions and under which premises?
- What are the technical risks?

Based on the result of the feasibility analysis a decision is made whether or not to continue the project (for further evaluations see BP5). The results are documented.

The risk analysis (see MAN.5 or, for hazard analysis, see the excursus at ENG.4 BP3) is an instrument used as a precaution against damage and is typically done in the form of a risk workshop during which further aspects are also considered, for instance, risks arising from software requirements. The focus is on technical risks, whereby in small projects other (non-technical) risks are usually identified at the same time. Large projects usually conduct several risk workshops, the results of which need to be consolidated by project management.

Regarding testability, the analysis needs to consider if a requirement is really testable at all or what the effort would be. The main basis for an analysis regarding testability are the verification criteria which specify what needs to be fulfilled in order for a requirement to be considered successfully verified. Verification criteria[18] are developed within the scope of BP2. This requirement results from the third note[19] of the ENG.2 process outcomes.

---

18. See excursus »Verification Criteria« in section 2.24.

The evalation of testability does require a first grasp of the system architecture and other system characteristics and cannot be done based on requirements alone. As a result of the analysis valuable indications are obtained not only with respect to the specification of the system requirements but also perhaps with respect to a system design optimized for testability. Evaluating testability at this time therefore makes a lot of sense because testability determines later test effort and the degree to which a system is testable. The test effort depends on further factors, such as quality requirements and the test environment to be used. Regarding requirements, the test effort is affected by:

- Redundant requirements that cause redundant tests
- Unclear requirements that hamper the definition of test cases or make it impossible
- Non-testable requirements without quantifiable criteria
- Missing requirements, for example because interfaces are not addressed

In the evaluation of testability it is useful to assess the requirements according to their testability (e.g., A = »testable«, B = »testable under certain conditions«, C = »not testable«). This evaluation can be made during reviews. It also makes sense during the evaluation to create a first version of the test plan[20].

BP3: Determine the impact on the operating environment. *Determine the interfaces between the system requirements and other components of the operating environment, and the impact that the requirements will have.*

System requirements analysis also considers the system's direct operating environment and operating conditions. The identified system requirements as well as the results of the feasibility, risk, and testability analysis may have an impact on other systems in the direct environment. Since they constitute the system's direct operating environment they need to be considered. Example: The ECU can check the status of the engine immobilizer prior to engine start. An interesting question, especially with regard to highly networked control units, is what kind of behavior the system is going to show under certain operating conditions, e.g., a possible malfunction in case of implausible bus messages. The effects on the operating environment can, for instance, be determined during the risk analysis (see BP2) or in the form of a »hazard and operability study« (HAZOP, see excursus in ENG.4 BP3) at system level. The HAZOP results are then taken into account in further development.

Requirements may also have a reciprocal effect with the system's operating conditions. To these belong, for instance, climatic conditions, fuel quality, the system's mounting location, etc.

---

19. The third note states: »Analysis of system requirements for testability includes development of verification criteria.«
20. See excursus »Test Documentation« in ENG.6.

2.4 ENG.2 System Requirements Analysis

BP4: Prioritize and categorize system requirements. Prioritize and categorize the identified and analyzed system requirements and map them to future releases of the system.

NOTE: *Refer to the process SPL.2 Product Release.*

Prioritization and categorization of system requirements form the basis for release planning (Level 2) where requirements are allocated to individual releases and deliveries. Requirements prioritization is a basis for adjusting the scope of the work and for project planning. Low-priority requirements will perhaps be dropped when faced with resource shortages or postponed to later releases[21]. A prioritization of system requirements can, for instance, be done based on the following criteria:

- Rapid prototyping
- Rapid implementation of basic functionality
- On-schedule deliveries as agreed with the customer and on-schedule SOP
- Provision of interfaces to other systems or components at defined dates
- Cost-benefit consideration
- Customer wishes

Categorization should primarily be done according to system components (software, hardware, mechanics) or according to subprojects or teams.

BP5: Evaluate and update system requirements. Evaluate system requirements and changes to the customer's requirements baseline in terms of cost, schedule and technical impact. Approve the system requirements and all changes to them and update the system requirements specification.

The technical feasibility assessment of system requirements (feasibility, risk, and testability) is already required in BP2. BP5 additionally requires the exploration of the technical impact. For instance, if a safety related system is being developed, a detailed consideration of the requirements with respect to their impact on the functional safety must be carried out.

Other supplier departments, such as the controlling, management or sales departments, also carry out an evaluation of the system requirements. For instance, in an early phase of the project they may use the project manager's (see MAN.3) system requirements based cost estimate to evaluate if the project is economically feasible. Together with the evaluation from BP2 a decision can be made whether or not to continue with the project; for instance, whether a quotation is to be issued and if so, what the price will be.

---

21. Release planning specifies when or in which product version particular features will be implemented. Based on this, project management can structure development and prioritize work (see SPL.2).

If, during the course of the project, changes to system requirements or new system requirements are proposed, a controlled workflow must be in place regulating analysis, approval, rejection, and updating of respective points in the system requirement specification (similar to what is described in ENG.1 BP5).

Usually, at the beginning, the project manager is in charge of any required approval while later on changes are approved by a steering committee or Change Request Board (CRB, see SUP.10 BP6).

BP6: Ensure consistency and bilateral traceability of customer requirements to system requirements. Ensure consistency of customer requirements to system requirements including verification criteria. Consistency is supported by establishing and maintaining bilateral traceability between the customer's requirements and system requirements including verification criteria.

The customer requirements identified in ENG.1 are the basis for the system requirements analysis. In this context, consistency means that customer requirements and system requirements are consistent. In order to judge consistency at all, we must know which system requirement is connected content-wise with which customer requirement, and vice versa (= bidirectional traceability). However, in most cases traceability exists for only part of the system requirements, since many system requirements are based on other sources (e.g., internal requirements derived from product policies and product architecture or platform strategies). Since these sources may also be considered to be from a customer source, traceability must be established here, too.

The request for traceability also extends to the verification criteria[22] included in the system requirements. Consistency checking with customer requirements is best done in a review.

BP7: Communicate system requirements. Establish communication mechanisms for dissemination of system requirements, and updates to requirements to all relevant parties.

The required mechanism must ensure that requirement and change-related information is provided unprompted to all project members who need it for their work. First of all, those who actually need this information must be identified. Here, too, the creation of a communication plan (see figure 3–5) turned out to be useful, stating which roles are to be informed generally and in case of changes. The role responsible for this information actually being communicated must also be specified. This requirement is included in similar form in the ENG.2–ENG.4 processes.

---

22. As stated in the second note of the ENG.2 process outcomes, verification criteria are typically included in the system requirements. However, verification criteria can also be included in the test plan; see the excursus on »Test Documentation« in ENG.6.

2.4 ENG.2 System Requirements Analysis

The method by which this communication is frequently implemented is that requirements documents and change notes are stored at a unique location in the file or configuration management system and accessible to all involved parties. It is also important to establish an appropriate infrastructure for the publication of information such as system requirements. For this purpose one may also use database solutions with HTML GUIs that are integrated in the respective company intranets. Changes discussed in project meetings can additionally be distributed via e-mail (for instance, via meeting minutes).

> **Note to Assessors**
>
> The evidence that these mechanisms work lies in the proof in relevant tools and systems or in correspondence using a suitable distribution list.

2.4.4 Designated Work Products

13-22 Traceability Record

This record is used to show the (bidirectional) links between requirements, design elements, and code, etc., via the individual development phases. Mapping tables or traceability matrices are commonly used in practice to show these connections (see also section 2.24).

> **Note to Assessors**
>
> Automotive SPICE describes that requirements traceability and its implementation over different phases of the lifecycle should be established in both directions (bidirectional) and should therefore be possible at every point of the lifecycle model and in both directions. In the assessment, individual work products and their links to previous and subsequent work products are sample-checked across different development phases.

17-08 Interface Requirements

Requirements on interfaces typically refer to data formats, timing, call hierarchy, interrupts, etc. A distinction is made between internal and external interfaces. Internal interfaces describe the interaction between system elements. External interfaces exist to other systems in the operating environment and to users.

17-12 System Requirements

System requirements are collected or derived from customer requirements documents (e.g., customer requirements specification) and written down in a system requirements document (e.g., the system requirements specification). A system requirement describes functions and capabilities of a system. Requirements categories are, for example, functionality, requirements on the part of the organization, safety aspects, ergonomics, interfaces, operational and maintenance requirements, and design constraints. In addition, the system requirements document provides an overview of the entire system and of the relationships of its components including, especially, those system elements with software.

Design suggestions and quality criteria for this type of document are, for instance, provided in IEEE 830-1998 (Recommended practice for software requirement specifications), which can also be applied to system requirements ([IEEE 830], see also figure 2–4).

2.4.5 Characteristics of Level 2

On Performance Management

The analysis of the system requirements originating from customer requirements must be done in a systematic way and requires appropriate planning. Planning of the process includes, for instance, scheduling of requirements baselines and requirements reviews. Updating of system requirements after changes to customer requirements is in most cases asynchronous. Agreed changes need to be considered in schedule and resource planning.

On Work Product Management

The requirements of Process Attribute PA 2.2, apply particularly to the system requirements documentation (e.g., the system requirements specification). This includes, for example, that system requirements are subject to review and should be accepted or confirmed by the customer. Baselines contain the versions of the customer and supplier's requirements documents that were valid at the time.

2.5 ENG.3 System Architectural Design

2.5.1 Purpose

The purpose of the system architectural design process is to identify which system requirements are to be allocated to which elements of the system.

At the top level, system architecture describes all system elements, their relation to each other, and interfaces. For each system element the associated system requirements must be known.

2.5.2 Characteristics Particular to the Automotive Industry

During the definition of the system architecture design a decision is made as to which functions will be implemented in the hardware and which will be implemented in the software. For example, a decision must be made whether to implement a display function using the software control function of an already existing LCD display or a separate LED.

In complex systems, the division of a system into its elements is often done in several steps or description levels. In those cases the description consists of several related documents with differing degrees of detail. Figure 2–5 provides an example of a system architecture design in the early stages. Figure 2–6 contains an example of a lower description level.

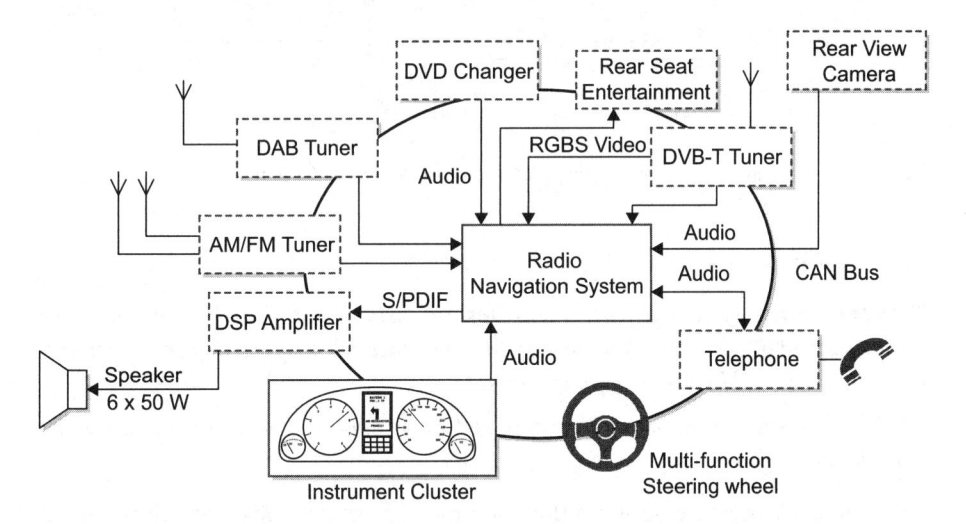

Figure 2–5 Example of a top-level system architecture design for a radio navigation system

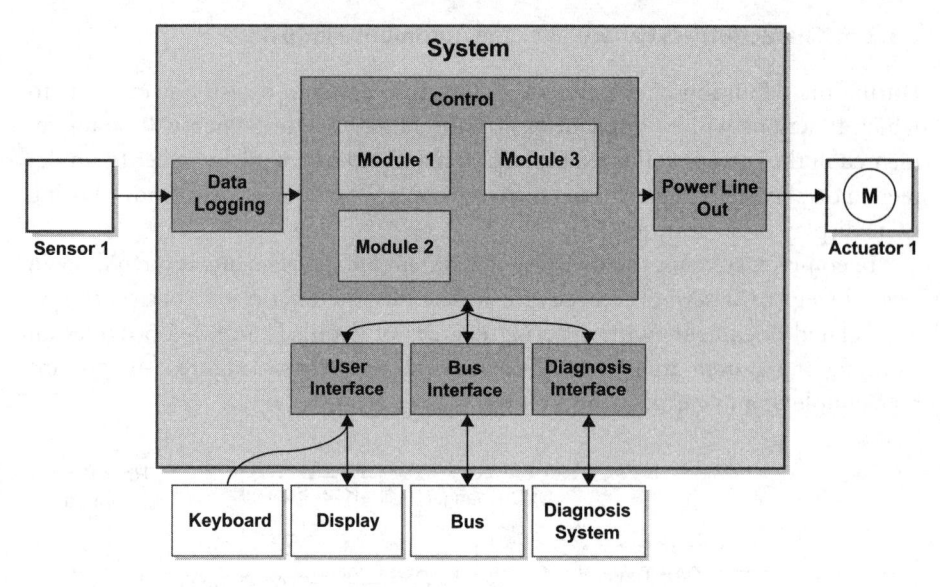

Figure 2–6 *Example of a low level system architecture design description*

2.5.3 Base Practices

BP1: *Define the system architectural design.* *Establish the system architecture design that identifies the elements of the system with respect to the functional and non-functional system requirements.*

NOTE: *The system might be decomposed into several subsystems on different system levels, if necessary.*

The system architecture design is the uppermost, most generic description level in the form of overview pictures, lists, and descriptions of system elements. It consists of elements that have been identified, in correspondence with the system requirements, to be implemented. During system architecture design, the corresponding verification criteria[23] need also to be developed. This requirement results from a note to the process outcomes in ENG.3[24]. Verification criteria specify what needs to be fulfilled in order to be able to consider the system architecture design–primarily through integration testing–to be successfully verified.

---

23. On this point, see the excursus »Verification Criteria« in section 2.24.
24. The corresponding requirement reads: »Definition of system architectural design includes development of verification criteria.«

BP2: Allocate System Requirements. Allocate all system requirements to the elements of the system architectural design.

Identified functional and nonfunctional system requirements are—as far as possible—assigned to system architecture design components. The objective is to try and assign all requirements to ensure faultless interaction of all system elements in due consideration of all constraints, and to avoid iteration loops. In complex systems, which consist of many interrelated subsystems or system elements, such assignment is not always possible at that stage. The typical reason for this is that at this early point in time detailed design decisions have not yet been made and that no decision can be made as yet regarding the relevance of the requirements in question. Assignment must then take place at a later date. Some (especially nonfunctional) requirements apply to the system as a whole, so that an assignment to individual system components is not possible.

Assignment can, for instance, be achieved by structuring the system requirements description (e.g., the system specification) having the system architecture design already in mind. The result of the assignment is traceability between system requirements and system architecture design.

BP3: Define interfaces. Identify, develop and document the internal and external interfaces of each system element.

At the top level a definition and description is given of how the system elements interact and how the system functions in its interaction with other systems in order to realize the required system functions (identified in ENG.2).

In doing so, a distinction is made between internal and external interfaces. Internal interfaces describe the interactions between individual system elements. Interface descriptions can either be documented in the system architecture design or in separate documents. They are often used for the description of certain specifics, for instance, those that are to be considered while coding hardware-related layers or when existing drivers or operating systems are used.

External interfaces are the interfaces of the system to other systems, e.g., to data transmission, displays and control elements, or sensors/actuators.

BP4: Verify system architectural design. Ensure that the system architecture meets all system requirements.

Checking the system architecture against the system requirements can assume different forms. Besides coordination meetings involving different groups, it is recommended that an architecture review is performed and that the results are documented. In practice it may be difficult at first to map all the factors to be considered onto the system requirements. Across-the-board cooperation is, therefore, important.

Coordination meetings are conducive to drastically reducing the risk of conceptual and, therefore, probably more cost-intensive changes at a later stage in the project. The need for coordination exists not only internally but also externally to external partners. Internal cooperation is to be agreed upon with all individuals internally involved in the project. These include the following:

- The purchasing department, since, for instance, third-party system components (e.g., standard software, electronic components) must be agreed upon and acquired. A need for cooperation exists, especially regarding the standardization of third-party acquisitions, safeguarding delivery reliability, etc.
- Domain representatives to coordinate strategic development issues
- Representatives from other or parallel projects from which elements will be taken over in the course of platform developments
- Designers, developers, and testers who will have to work with the system architecture design

External coordination is necessary in the following instances:

- If external system interfaces must be explicitly specified and coordinated with the customer because the system under development will typically be integrated into an environment consisting of other (possibly coequal) systems,
- If the customer has a say in the design of the system architecture because he specifies or prohibits particular system components (e.g., operating system or specific drivers).

BP5: Ensure consistency and bilateral traceability of system requirements to system architectural design. *Ensure consistency of system requirements including verification criteria to system architectural design including verification criteria. Consistency is supported by establishing and maintaining bilateral traceability between the system requirements including verification criteria and system architectural design including verification criteria.*

The basis of this check is the traceability between system requirements and system architecture design; see BP2 on this point. It must be ascertained that each system requirement has been considered in the system architecture design and that for each element of the system architecture the relevant system requirements are known, and that the system requirements and the system architecture design are consistent. During the project care must be taken that full traceability is preserved. For this reason no requirement-relevant changes may be made on the system architecture design without repeating relevant steps of the system requirements analysis, and vice versa. Where design requirements are not based on system requirements it is desirable (but not mandatory) to document their origin and necessity.

The requirement for traceability extends also to the system architecture verification criteria[25]. Consistency with system requirements is best checked during

---

25. Verification criteria may be included in the test plan, see the excursus on »Test Documentation« in ENG.6. Otherwise see the excursus on »Verification Criteria« in section 2.24.

a review. Special care should be taken to check that both types of verification criteria are consistent in order to coordinate system and system integration test effectively, and to ensure consistency between the two.

BP6: Communicate system architecture design. Establish communication mechanisms for dissemination of the system architectural design to all relevant parties.

This is to ensure that engineers working on system architecture design or affected by changes to it are kept duly informed and are able to react to, or consider, such modifications in their own work. The statements already made in ENG.2 BP7 on communication mechanisms can be applied here, too.

2.5.4 Designated Work Products

04-06 System Architecture Design

System architecture design provides an overview of the structure of the system as a whole and describes the interaction of individual elements, including software. The overall system typically consists of one or several »block diagrams« illustrating the system elements, and an account of their interrelationships and data flows. These overview pictures are usually supplemented with technical descriptions (interface descriptions). In complex systems, the description may consist of several related documents varying in detail. Important in this case is the traceability of requirements/functions across several degrees of detail (see examples in figure 2–5 and figure 2–6).

13-22 Traceability Record

See ENG.2

13-25 Verification Results

The verification results of the system architecture design against the system requirements (see BP4) are manifested in the form of review or meeting support records, completed checklists, and verification logs (for further explanations related to this work product see SUP.2).

2.5.5 Characteristics of Level 2

On Performance Management

A system architecture must be designed systematically. In practice, small work steps, repeated in many iterations, are rarely planned in detail. Larger work steps (e.g., milestones indicating when different versions of the system architecture design need to be available), however, must definitely be planned and documented. If, at certain milestones, the need arises to update the system architecture design, it must be planned in the project plan.

On Work Product Management

The requirements of Process Attribute PA 2.2 are particularly relevant for the system architecture. They require, among others, a review of the system architecture and, as the case may be, acceptance by the customer. Baselines include the system architecture relevant in each case.

2.6 ENG.4 Software Requirements Analysis

2.6.1 Purpose

The purpose of the software requirements analysis process is to establish the software requirements for the system.

In ENG.3, the hardware and software system elements were identified. Now, from this process onward up to ENG.8, consideration is given exclusively to software parts that form the system's software items. The software requirements analysis is the intermediate step between ENG.3 system architecture design and ENG.5 software design. In ENG.4, the requirements for the software items are identified. Software requirements are categorized into functional and nonfunctional requirements.

2.6.2 Characteristics Particular to the Automotive Industry

In practice, the transitions between the processes ENG.2 to ENG.5 are mostly blurred and are by nature iterative and recursive. The code must not only satisfy functional requirements but also other, nonfunctional requirements identified within the scope of the software requirements analysis. Besides the requirement for compliance with coding guidelines such as MISRA Rules [MISRA], additional quality requirements (e.g., metrics) that positively affect the quality of the software are specified for the source code. Examples are analyzability, modifiability, stability, and testability.

More and more formal methods used for software requirements analysis have been applied in recent years. Formalization of the requirements identification, not only for systems but also for software can, for instance, be accomplished through the creation of use case diagrams. Use-case diagrams are graphical and textual representations of functionality using UML[26] as the description language. The advantages of this formal procedure manifest themselves in terms

---

26. Unified Modeling Language; UML™ is a graphical description method for object oriented models standardized by the Object Management Group (OMG).

2.6 ENG.4 Software Requirements Analysis

of unambiguousness, better comprehensibility, effective communication of contents, independence from implementation, traceability, reusability, defect detection, and proof of correctness.

2.6.3 Base Practices

BP1: Identify software requirements. Use the system requirements and the system architectural design as the basis for identifying the functional and non-functional requirements of the software and document the software requirements in a software requirements specification.

NOTE: *In case of software development only, the system requirements and the system architectural design refers to a given operating environment (see also NOTE at BP3). In that case, customer requirements should be used as the basis for identifying the required functions and capabilities of the software.*

Functional and nonfunctional requirements on the software are identified and assigned to individual software items. Functional and nonfunctional software requirements can to some extent be taken over directly from the system requirement specification, others need to be transformed. Automotive SPICE explicitly requires that a software requirements specification is created.

> **Note to Assessors**
>
> Few projects establish a separate software requirements specification. Most describe all relevant requirements (functional and nonfunctional system and software requirements) in one single document. The reason for this is that although system functionality is mainly determined by software, it does not make sense to try and separate it from hardware functionality. Rating of this base practice should in such cases depend on a project's context and size. For complete achievement, evidence should be provided to the effect that the functional and nonfunctional software requirements have been clearly specified and that they adequately cover the functional scope. In other words, aspects of the system level and the software level must be visible, irrespective whether they are kept in one document or in several separate documents.

BP2: Analyze software requirements. Analyze the identified software requirements in terms of technical feasibility, risks and testability.

NOTE: *Verification criteria for all software requirements should be defined for further development of software test cases.*

NOTE: *The results of the analysis may be used for categorization of the requirements.*

The deliberations given in ENG.2 BP2 also apply to the analysis of software requirements. Software-specific risks can be:

- Use of insufficiently tested technical solutions or tools, e.g., for automatic code generation
- Designated development tool chains or test suites are incomplete
- nonfunctional software requirements cannot be met because code is auto-generated (e.g., the auto-generated code does not fit into the EEPROM).
- Increased test effort in case of auto-generated code

Here, too, an Automotive SPICE note requires the development of verification criteria as part of this base practice.

BP3: Determine the impact on the operating environment. *Determine the interfaces between the software requirements, system requirements and/or other components of the operating environment, and the impact that the requirements will have.*

NOTE: The operating environment is defined as the system that the software works in (e.g., hardware, operating system, etc.).

Our deliberations in ENG.2 BP3 also apply to this base practice but are limited to the direct operating environment of the software, i.e., the system component(s) on which the software runs. Software requirements have an impact on other software parts (e.g., the operating system) or on hardware in the direct environment (e.g., controller, memory). An important question one may ask is what kind of malfunctions the software may cause under certain operating conditions. To find out one can perform a situation analysis in the form of a »Hazard and Operability Study« (HAZOP) (see subsequent excursus),[27] the results of which are then considered in subsequent development.

Excursus: Example Method Hazard and Operability Study (HAZOP)

HAZOP was developed in the Seventies by the chemical industry, and extended in the Nineties to cover software development.

The objective of this method is to analyze the behavior of a software function through qualitative analysis (in a virtual environment) under different operating conditions, and to detect deviations from the target state. Moreover, it helps to detect and avoid hazards and severe malfunctions that might otherwise occur as a result of an unexpected event. Based on the HAZOP study, the software requirements or architecture may be accepted or rejected, resulting, in the latter case, in a revision of the design.

---

27. HAZOP is given here only as an example and is not required by Automotive SPICE.

2.6 ENG.4 Software Requirements Analysis

A HAZOP study is conducted by a team. A software function has so-called guidewords associated with it. For example, the software function »send message« is associated with the guidewords: too often, too early, too late, too little, and so on. The findings are collected in a table (figure 2–7). A guideword describes a hypothetical deviation from normally expected attributes. Based on these guidewords, causes of failures and their impacts are listed as deviations; these are then discussed and measures are proposed to minimize either the probability of the occurrence of the failure cause or its impact.

Guide-word	Deviation	Possible Cause	Consequences	Measures
Too often	High bus load	Non-compliance with the bus spec, calling party sending even if no one listens	Communication breakdown, messages get lost	1. Ensure that bus spec is known 2. Ensure that the bus-spec is understood 3. Verify through design review 4. Implement specific bus load tests
Too rarely	Faulty communication	Non-compliance with the bus spec, calling party not sending if someone listens	Communication breakdown, messages missing	see above

Figure 2–7   Example of a HAZOP table

BP4: Prioritize and categorize software requirements. *Prioritize and categorize the identified and analyzed software requirements and map them to future releases.*

NOTE: See also SPL.3—Product Release Process.

Our deliberations in ENG.2 BP4 also apply to this base practice. By now it should be evident from the release planning when or in which release software requirements are going to be implemented. Already at this point, software architecture issues are being included in the release planning because, besides functional prioritization, there also exists a technically reasonable implementation sequence for the requirements. Requirements associated with system or hardware related layers are typically implemented earlier since they form the basis for additional functions. Categorization is primarily based on the division into software architecture components or software subprojects.

BP5: Evaluate and update software requirements. Evaluate software requirements and changes to the system requirements and/or system architectural design in terms of cost, schedule and technical impact. Approve the software requirements and update the software requirements specification.

Our deliberations in ENG.2 BP5 also apply to this base practice, with the addition that modifications of the system architecture may also create a need for the evaluation and perhaps modifications of software requirements.

BP6: Ensure consistency and bilateral traceability of system requirements to software requirements. Ensure consistency of system requirements including verification criteria to software requirements including verification criteria. Consistency is supported by establishing and maintaining bilateral traceability between the system requirements including verification criteria and software requirements including verification criteria.

NOTE: *In case of software development only, the system requirements and system architectural design refer to a given operating environment (see also NOTE at BP3). In that case, consistency and bilateral traceability will be of customer requirements to software requirements.*

Our deliberations shown in ENG.2 BP6 and ENG.3 BP5 also apply to this base practice. Here, too, software requirements may be added which are not derived from system requirements but originate from other sources (e.g., internal requirements derived from product policy, software architecture, or software platform strategies). It is advisable to establish traceability to these sources, too.

BP7: Ensure consistency and bilateral traceability of system architectural design to software requirements. Ensure consistency of system architectural design including verification criteria to software requirements including verification criteria. Consistency is supported by establishing and maintaining bilateral traceability between the system architectural design including verification criteria and software requirements including verification criteria.

NOTE: *In case of software development only, see NOTE at BP 6.*

In analogy to BP6, consistency and traceability must be established between system architecture and software requirements. Regarding traceability, it must be clear which software requirement is associated with which system architecture component, and vice versa. This is trivial if the system architecture includes only one software component, but if it includes several software components the software requirements must be structured in a way that shows explicit assignment to each of these components.

BP8: Communicate software requirements. Establish communication mechanisms for dissemination of software requirements, and updates to requirements to all relevant parties.

Our deliberations shown in ENG.2 BP7 equally apply to this base practice.

2.6.4 Designated Work Products

13-22 Traceability Record
See ENG.2

17-11 Software Requirements
Software requirements are specifications for the creation of software that have a major impact on the quality and usability of the software. Among others, the following are considered: customer requirements, system requirements, system architecture, standards to be observed, restrictions/constraints, relationships of the software items with each other, performance characteristics, required software interfaces, safety characteristics, behavior in case of failures and recovery after failures (see figure 2–4 on the structure of a requirements document according to IEEE standard 830).

2.6.5 Characteristics of Level 2

The statements given in the corresponding section in ENG.2 also apply here.

2.7 ENG.5 Software Design

2.7.1 Purpose

The purpose of the software design process is to provide a design for the software that implements and can be verified against the software requirements.

Software design is used to illustrate how software requirements are to be implemented in code and what the components are that the software consists of. Function, mode of operation and interaction of the components are described. Software design takes software requirements as a basis and provides the specifications needed for coding. The design process often consists of several iterative steps in which, starting with software architecture design, the design becomes increasingly refined and detailed until detailed design is completed. There is an overlapping area between system architecture and software architecture because components of the software architecture may, for instance, be listed in the system architecture, and vice versa.

In many cases, the software to be developed is a combination of platform components (reused code) and new code. Frequently, reused design and code elements must meet increased requirements which can, for instance, be checked in design reviews. Design reviews can also be used to consider and answer design-relevant questions related to fault tolerance:

- Are failures detected, and how?
- How are failures classified?
- What is the system's behavior in case of a failure?

Stress tests performed later may provide information on the robustness of the design, for instance, regarding sensor malfunctions, cable breaks, startup behavior, or the sudden and sharp increase of interrupts. It only makes sense to perform hardware related tests during system testing (see ENG.10).

2.7.2 Characteristics Particular to the Automotive Industry

In the automotive industry, software development is often very low-level, i.e., hardware-related, which means other issues that go beyond pure function development must also be considered in the design. Examples are timing behavior (interrupts and time slots) and communication protocols. As far as timing behavior is concerned, the reactions and interactions of involved components such as processors, memory chips and bus controllers need to be considered. In doing so, possible states are described (e.g., from initialization and operation up to sleep mode). If needed, failure scenarios[28] are described, too. As far as communication behavior is concerned, it is primarily the sequence and contents of communication messages that are specified and described, subject to events. In practice, design description and design modeling (e.g., by means of state diagrams) are more and more tool based. This has several advantages. For one thing, if the tool supports this, formal criteria (such as design consistency) can be ensured, for another thing certain functions and states can already be simulated and evaluated prior to actual coding at the PC. More recent tools usually also support code generation.

2.7.3 Base Practices

***BP1: Develop software architectural design.** Use the functional and non functional software requirements to develop a software architecture that describes the top-level structure and all the software components including software components available for reuse.*

NOTE: See also REU.2—Reuse Program Management

---

28. Design FMEAs typically provide input for failure scenarios, considering and assessing the probability of failure occurrence, taking into account associated constraints.

2.7 ENG.5 Software Design

The requirements on software identified in ENG.4 are translated into a software architecture (also called top level design) as a description at the highest level, and documented. Here the necessary central design decisions are made. Ideally, software architecture components remain stable during development so that there is no need to modify them or their interfaces during the remaining development phases. This step is the forerunner to software detailed design (see BP6) and must be consistent with it.

In practice, several views of the software architecture are usually necessary for a complete mapping. These may, for instance, be the following:

- Structure view (architecture, used architectural patterns)
- Behavior/state view (sleep-mode, startup, shutdown, description of bootblock, monitoring techniques, etc.)
- Use-case view (implementation of the use cases in the architecture)
- Process view (task design, timing, memory layout)
- BIOS and services view (CAN and LIN drivers, hardware abstraction layer, watchdog, operating system integration, etc.)
- Call hierarchy, e.g., as a block diagram
- Resources view (planned RAM/ROM consumption)

If difficult or far-reaching architecture design decisions have to be made, it makes sense to submit several architecture proposals and to evaluate them according to the criteria mentioned. Once made, architecture decisions should be documented in a traceable way so that later on in the project the decision making process can always be reconstructed and further revisions avoided. Criteria that are becoming increasingly important are the reuse part of the architecture, as well as its suitability for future reuse.

BP2: Allocate software requirements. *Allocate all software requirements to the components of the software architectural design.*

All software requirements relevant to the software architecture are allocated to software components of the software architecture. Examples are requirements related to calibration, requirements for a startup task, and interface requirements at a high abstraction level.

Allocation can be done in the traceability matrix (see section 2.24) by simply listing or flagging corresponding software components. Alternatively, the software requirements specification can be structured according to the software architecture in preparation for detailed design to verify that all requirements have been covered.

BP3: Define interfaces. *Identify, develop and document the internal interfaces between the software components and external interfaces of the software components.*

The interface description consists of the definition of the information to be exchanged externally (e.g., with other systems, peripherals, users) and internally (between individual software components). In most cases, the BIOS and services view including the hardware abstraction layer and the integration of the operating system are also specified here. The runtime performance of the interface interactions is specified in BP4. The interface description should at least contain the following two points:

- A global map of the considered interface and involved software components, including BIOS and services view
- A list of data exchanged via the interface, and corresponding specifications (e.g., name, value range, communication protocol)

BP4: Describe dynamic behavior. *Evaluate and document the dynamic behavior of and interaction between software components.*

NOTE: Dynamic behavior is determined by operating modes (e.g., start-up, shutdown, normal mode, calibration, diagnosis, etc.), processes and process intercommunication, tasks, threads, time slices, interrupts, etc.

A software system is particularly characterized by its dynamic behavior. Hence, timing behavior and interactions of the software components are an essential part of the software architecture. If needed, the interface description provided by BP3 should be extended by the following items:

- A global map of the interface under consideration (e.g., by means of an interaction diagram) and all involved software components
- A description of the different operational modes (e.g., startup, shutdown, normal operation, calibration mode, diagnostic mode, monitoring techniques), including all processes and inter-process communication, as well as tasks or threads necessary for operation
- A description of the interaction, e.g., timing behavior, call hierarchy, interrupts.

The description of the dynamic behavior of software can also be done by means of a dynamic model, whereby object changes and their relationships to each other are considered over time. Relevant components are:

- Control flow
- Interactions of objects
- Control of processes of a software system's simultaneously active objects

2.7 ENG.5 Software Design

During program execution, individual objects react to events, perform corresponding actions, and change their state. A dynamically working object can be described by the terms »state–event–condition–action–subsequent state«. The entire dynamic model of software systems consists of any number of such objects, whereby the model shows the sum of the activities for an entire system. All dynamic objects work in parallel and can change their states independently from each other. This dynamic modeling is independent of its implementation on the target system.

There are tools available on the market that support this approach, covering the areas of design, coding, test, and documentation. Dynamic objects are represented as finite state machines; states, events, and actions can be created at will and may be modified further for subsequent work steps (e.g., simulations). There are always different ways to describe objects or states thus created (e.g., in the form of a state diagram). Integrated simulators can create sequence diagrams of the dynamic behavior already in the design stadium.

BP5: Define resource consumption objectives. *Determine and document the resource consumption objectives for all software components.*

NOTE: *Resource consumption is typically determined for resources like Memory (ROM, RAM, external / internal EEPROM), CPU load, etc.*

In the automotive industry, resources are usually very limited and a relevant cost constraint. It is therefore vital to plan, measure and document early how targets for resource consumption, i.e., load factors for ROM, RAM, external/internal EEPROM, and perhaps CPU load[29], change during the development phases. Despite cost pressure, it may be a good idea to reserve a certain buffer of resources to be able to deal with late changes towards the end of development. The software architecture should be extended by at least the following points:

- Overview description of the planned and used resources of the defined software components for each development phase
- Description of the component's memory layout

BP6: Develop detailed design. *Decompose the software architectural design into a detailed design for each software component describing all software units and their interfaces.*

NOTE: *Task execution time is highly depended on target[30] and loads on the target which should be considered and documented.*

---

29. In telematics, a CPU load of 100% is not unusual. In most cases, response times and target achievement percentage are defined as a result.
30. This refers to the target ECU.

Based on the software architecture, detailed design is often developed iteratively. In complex systems this usually means several description levels. The goal is to arrive at a well-specified design that is sufficiently precise to be coded or code-generated in the subsequent implementation phase.

Software detailed design includes all the detailed functions and algorithms to be implemented, all input and output values, used data structures and, if required, resource consumption (for instance, storage allocation). There is often a need for clarification and increased accuracy regarding requirements during this phase, and often weaknesses and inconsistencies are detected regarding requirements.

BP7: Develop Verification Criteria. *Define the verification criteria for each component concerning their dynamic behavior, interfaces and resource consumption based on the software architectural design.*

See section 2.24, excursus »Verification Criteria«. Verification criteria are the basis for the specification of the subsequent tests, especially for ENG.7.

BP8: Verify software design. *Ensure that the software design meets all software requirements.*

Usually, a design review is performed to check the design for correctness and testability. Other means to check testability are test planning and the creation of test cases. Typically, the following evaluation criteria play a role regarding the testability of software components:

- Complexity of the software design[31] (number and unit size, call hierarchy, dependencies, recursions, use of variables, etc.)
- Definitions regarding interfaces and communication
- Software design modularity, encapsulation

On testability, see also ENG.2 BP2. The software design is revised depending on the result of the design reviews. In case of larger changes the design review must be repeated. Review results and decisions are documented.

BP9: Ensure consistency and bilateral traceability of software requirements to software architectural design. *Ensure consistency of software requirements including verification criteria to software architectural design including verification criteria. Consistency is supported by establishing and maintaining bilateral traceability between the software requirements including verification criteria and software architectural design including verification criteria.*

Consistency, in this context, means consistency between software requirements and software architecture. This presupposes that the software requirements (if relevant) have been correctly implemented in the software architecture and that

---

31. Quantifiable, for instance, with a McCabe complexity metric.

all relevant software requirements have been considered in the software architecture.

In order to judge consistency we must know which software requirements have found their way into the respective architecture elements. Consistency checking can be done partly while setting up traceability, although the primary means is still the design review already described in BP8. On the whole, part of the »vertical« traceability is created here. The statements made in section 2.24 »Traceability in Automotive SPICE« apply here also.

BP10: Ensure consistency and bilateral traceability of software architectural design to software detailed design. *Ensure consistency of software architectural design including verification criteria to software detailed design including verification criteria. Consistency is supported by establishing and maintaining bilateral traceability between the software architectural design including verification criteria and software detailed design including verification criteria.*

Similar to BP9, consistency here defines consistency between software architecture and software detailed design. Again, the primary means for checking consistency is the design review already described in BP8. The implementation of traceability, as required in BP10, completes the »vertical« traceability. Now all software requirements can be vertically and bilaterally traced via the software architecture right down to software detailed design. The statements made in section 2.24 »Traceability in Automotive SPICE« also apply here.

2.7.4 Designated Work Products

04-04 High Level Software Design[32]

The software architecture provides an overview of the structure of the software system as a whole and describes the interaction of the different components. It usually consists of one or several block diagrams illustrating the software components with their interrelationships (see figure 2–8). These overview diagrams are often supplemented with technical descriptions (e.g., interface descriptions). In complex software systems, the description may consist of several, related documents with varying degrees of detail.

---

32. To ensure consistency, we will use the term »software architecture« as introduced in BP1.

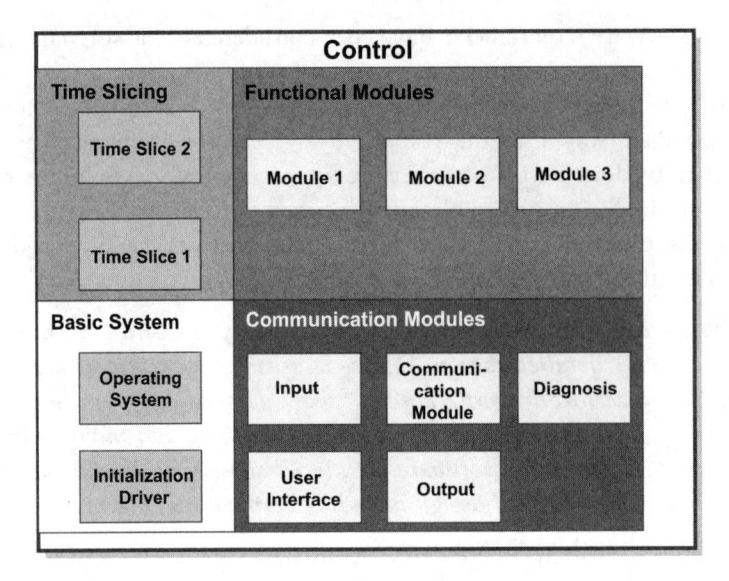

Figure 2–8   Example of a top-level software architecture

04-05 Low Level Software Design [33]

Software detailed design provides the detailed specification of the software items, their interactions, interface descriptions (input and output data), algorithms, assignment of memory space, data specification, and specifications concerning the program structure. Besides descriptions in natural language, a variety of other forms is commonly used. These include flow charts, symbolic programming languages, finite state machines, state charts, message sequence charts, or data relationship models. Furthermore, software detailed design includes the definition of naming conventions, formats of required data structures, data fields, and the purpose of each required data element. In practice, software detailed design is often generated with tool support and the objective is to later generate source code automatically. Modeling and simulation of individual functions is done on a PC. This way, for instance, finite state machines are diagrammed with a graphical editor. Allowed state transitions are specified by events and conditions. Created models can be integrated in hardware-in-the-loop or rapid control prototyping systems for further modeling operations.

13-22 Traceability Record

See ENG.2

---

33.   To ensure consistency, we will use the term »detailed design« as introduced in BP6.

2.7.5 Characteristics of Level 2

On Performance Management

The creation of the software design must be done systematically. In practice, »small« work steps that will be repeated in many iterations are rarely planned in detail. However, »larger« work steps (e.g., design reviews and milestones at which different software design versions are expected to be available) certainly need to be planned and should be documented in documents like the project plan. If larger updates of the software design can be foreseen at certain milestones, they need to be included in the project plan.

On Work Product Management

The requirements of Process Attribute PA 2.2 particularly apply to software design documents. They include conducting a design review and, as the case may be, acceptance of the software design by the customer. Baselines include the design documents that are valid in each case.

2.8 ENG.6 Software Construction

2.8.1 Purpose

The purpose of the software construction process is to produce verified software units that properly reflect the software design.

Based on the software design created in ENG.5, the software is now being developed. In practice, the coding process proceeds iteratively in several successive coding, defect detection, and defect correction cycles and is strongly interconnected with other development activities, such as software design or unit testing[34].

2.8.2 Characteristics Particular to the Automotive Industry

In the automotive industry, requirements of the hardware used are very much taken into account during coding. Especially in hardware-related layers like operating systems or device drivers, generated code can often be used for one micro controller only, or at most for one controller family. The same applies to code generation tools (e.g., compilers) that are not universally applicable. Common programming languages are C, assembler, or C++ for non-hardware-related layers. If software modeling, simulation and design description is tool-supported, automatic code generation is usually triggered in a subsequent work step. How-

---

34. Unit testing is not to be confused with defect detection during development (debugging), see BP4.

ever, the requirements of the project are often only partially satisfied by automatically generated code. Execution times are often suboptimal and autocode memory capacity requirements frequently exceed available memory capacity. In most cases, memory costs are considerable. In practice, therefore, we very often see further optimization efforts to reduce memory requirements. Individual enterprises differ profoundly in the ways and methods they use for code generation. On the one hand, they may range from complete tool chains that comprise the entire software development process from requirements definition right down to software test and support activities (e.g., configuration management). On the other, they may comprise completely heterogeneous structures with considerable gaps in the tool chain and some tools having been developed internally.

> **Note to Assessors**
>
> In principle, tools are not rating-relevant in Level 1 and 2 assessments (e.g., how good they are), since Automotive SPICE does not make any demands in this respect (explicit requirements regarding adequate infrastructure exist at Level 3). However, the question as to whether their application adversely affects the performance of base practices is indeed very relevant for rating purposes. For this reason one may, for instance, check if a tool is appropriately deployed and administrated (e.g., access rights of a CM tool), if the entry and exit conditions for tool application in the project are defined, i.e., the transition from »manual activity« to tool application and vice versa, and if relevant specifications were followed.

2.8.3 Base Practices

BP1: Define a unit verification strategy. Develop a strategy for verification and re-verifying the software units. The strategy should define how to achieve the desired quality with the available techniques.

NOTE: Possible techniques are static/dynamic analysis, code inspection/review, white/black box testing, code coverage[35], etc.

The verification strategy[36] addressed here is part of a overall verification strategy for the ENG.6 to ENG.10 processes, whereby in each of these processes a similar strategy is required. This overall verification strategy (see also SUP.2 BP1 and work product 19-10 »Verification strategy«) is supposed to accomplish the harmonization and coordination of the different methods (reviews, tests ...) and verification activities during the different development phases (coding, integration, software test, and so on). The reason for this need for coordination is that the

---

35. Code coverage analysis, see explanation in the glossary.
36. A »strategy« is a long-term oriented, tactical pursuit of a goal. It may, for instance, be stated in a project description or in a plan. However, the presence of a plan is only required at Level 2 by Process Attribute PA 2.1.

2.8 ENG.6 Software Construction

different verification methods simultaneously overlap and supplement each other. In unit testing, for instance, defects can be found which are not detected in code reviews, and vice versa. Apart from this there is an overlapping area between the two methods. Something similar holds true for the coordination of the different test levels. For example, the classical distinction between unit test, integration test, and software test is based on the concept that unit testing aims at securing the functionality of the individual units as measured by the software design, that integration testing aims at testing the interfaces and interactions as measured by the software architecture, and that software testing verifies the overall functionality as measured by the software requirements. In the automotive industry, this classical division of work is often far less stringently adhered to in smaller systems where tests are often combined to make up a joint test because it saves effort and is more practicable. From the Automotive SPICE's point of view, this is perfectly legitimate as long as the purposes of the processes in question are fulfilled.

> **Note to Assessors**
>
> It is not necessary to implement 1:1 each of the Automotive SPICE processes separately. An organization needs to design its processes in such a way as to suit its own particular circumstances. The task of an assessment team is to establish the correspondence between the organization's processes and the requirements set by Automotive SPICE, and to assess the fulfillment of the latter. Therefore, where test levels are combined, the particular characteristics associated with different test levels will still have to be taken into account. If an organization fails to do so it will be downgraded regardless of its size.

Reverification is another issue for which a sound concept is needed. The term reverification denotes renewed verification after a software unit has been modified. There are explicit demands for a »regression test strategy« (see also our deliberations shown in ENG.7 BP8) in the ENG.7 to ENG.10 processes. In ENG.6, however, we find other possible reverification methods besides regression testing so that there is a need to agree on the following question: Which method of verification is used for which type of change?

Verification is always performed against specifications that typically come from the previous development phase. In the process purpose, Automotive SPICE requires verification against the design. For code, additional standard practices apply, particularly with regard to documentation and coding guidelines (e.g., MISRA rules[37]). Further specifications result from the strategy mentioned in BP1.

Moreover, BP1 talks about the »desired quality«. Here the thing to do may be to agree on different quality levels for the different types of deliveries to the

---

37. Although only stated as a possibility in Automotive SPICE, most OEM's require compliance with MISRA rules.

customer. Some OEMs, for instance, require the delivery of software versions within very short intervals (in extreme cases in weekly increments). It is obvious that these increments cannot be of the same quality as later ones close to production. Other possible quality requirements (not necessarily at the level of the individual software unit) are:

- Software integrity levels according to [IEEE 1012] for the specification of the criticality of a system part regarding the system as a whole; such levels can provide an important contribution towards effort optimization regarding verification activities in that critical system components are verified more intensively than less critical ones.
- Safety integrity levels according to [IEC 61508], which also affect the verification method and verification intensity

BP1 states the following possible verification methods:

- Static analysis (see also glossary): During static code analysis[38], code is checked for consistency and compliance with rules and conventions with the help of tools. In assessments, OEMs often view test records to see if the MISRA rules that were already mentioned have been observed.
- Dynamic analysis (see also glossary): These comprise all verification methods in which code is being executed, in particular different types of tests. Unit tests are typically performed as white-box tests. Besides verifying functionality, these tests are also supposed to take into consideration—as far as possible in unit testing—timing behavior, communication and the fulfillment of non-functional requirements (see also our excursus on »Test Methods« in ENG.8).
- Code inspection and code review (see also glossary): These verification techniques exist in many different variants. Here, code is typically analyzed in advance by one or several domain experts, and the results are then discussed in a review meeting. Typical issues to be checked are program logic and adherence to coding guidelines. Often checklists are used that include critical issues and which were derived from major problems found in earlier projects.
- White-box tests and black-box tests (see also glossary): see our excursus on »Test Methods« in ENG.8.
- Code coverage analysis (see glossary)

BP2: Develop unit verification criteria. *Develop and document criteria for verifying that each software unit satisfies its design, functional and non-functional requirements in the verification strategy.*

NOTE: *The verification criteria should include unit test cases, unit test data, coding standards and coverage goals.*

---

38. These belong to the White Box test group.

NOTE: The coding standards should include the usage of MISRA rules and defined coding guidelines.

Verification criteria specify what needs to be satisfied for the software unit to be considered successfully verified. Verification criteria for a software unit are more comprehensive than the verification criteria referenced in BP5-7, since the latter relate to only one requirement each (for further details, see section 2.24, excursus »Verification Criteria«). BP2 specifies in its notes some typical verification criteria:

- Unit test cases that are to be successfully executed; on this point, see the definition of »Test case« in the excursus »Test Documentation« at the end of this section.
- Unit test data: These refer to the input values for a test object and the expected results and their interrelationships with each other (e.g., timing), as well as the target values after execution of the respective test case, including tolerances, response time, and so on. This data is part of the description of a test case (see above).
- Coding rules and guidelines: MISRA rules were already addressed in BP1. Coding guidelines typically define rules concerning file naming conventions, file organization, commenting, layout, naming conventions, declarations.

BP3: Develop software units. *Develop and document the executable representations of each software unit.*

NOTE: In the development of software units code generation tools can be used to reduce the manual coding effort.

The software units are coded, defects are detected[39] and removed until the developer is satisfied that the code meets the required performance features. If the functional scope is expanded in several iterative steps, the cycle of coding, defect detection and defect removal is also performed several times.

It is important that regulations like coding guidelines or other nonfunctional requirements are being observed during programming. Later readability and understandability, and perhaps reusability of the code, depend on the observance of these regulations. The same applies to stability, testability, changeability, and maintainability.

As a minimum, documentation is done in the code itself (comments, explanations, change history) and, if required, in other documents, too (e.g., »release notes«). More generally speaking, design documentation and user documentation (installation, operating and maintenance documentation) can also be considered to belong to this group.

---

39. Mostly referred to as »debugging« and supported by development tools that allow halting the execution at certain breakpoints, for instance, to look at variable values. On the separate issue of unit tests see BP2 and BP4.

> **Note to Assessors**
>
> Important for code quality are the quality and understandability of the work results from previous development phases, i.e., the requirements and design documents. It is of benefit if developers have already become involved in the requirements and design phase. Otherwise the information flow or know-how transfer must be ensured in a different way. This should be checked. Moreover, it should be checked whether the development organization has created appropriate technical prerequisites for a practicable and efficient code generation (infrastructure, tool use, etc.).

BP4: Verify software units. Verify software units against the detailed design according to the verification strategy.

Created code is checked according to the verification strategy, and detected defects are removed. The goal of verification is to prove with reasonable effort that the software meets its specifications, and that risks related to the correct function of the software units are kept within reasonable limits.[40]

> **Note to Assessors**
>
> The existence of a separate test level for unit tests is not mandatory if their purpose is met by combined test levels; see also BP1.

BP5: Record the results of unit verification. Document the results of unit verification and communicate to all relevant parties.

As documentation, we expect at least a test log as well as test incident reports (see the excursus »Test Documentation« at the end of this section). The results are usually communicated within the team, e.g., to individuals whose task it is to remove the defects, to the person in charge of integration, and to the (sub)project manager.

---

40. In practice there is no such thing as 0-defect software, and one cannot prove 0-defect software with verification methods, but only reduce the risks. The only way to really prove absolute freedom from defects is via formal verification of the software. Unfortunately, the cost-benefit ratio of this elaborate method is very poor and even if the software were to be formally verified the compiler, operating system and hardware would have to be examined in the same way to really guarantee freedom from defects. For these reasons, formal verification is very rarely used in the automotive industry and only where central, safety-critical algorithms are involved.

2.8 ENG.6 Software Construction

BP6: Ensure consistency and bilateral traceability of software detailed design to software units. Ensure consistency of software detailed design including verification criteria to software units including verification criteria. Consistency is supported by establishing and maintaining bilateral traceability between the software detailed design including verification criteria and software units including verification criteria.

Consistency of each developed software unit with the software design must be ensured. A consistency check must ensure that:

- There is no software unit that is not described in the software design; and vice versa, for each design element there exists an associated software unit.
- Each software unit reflects the design specifications.
- Verification criteria of the software design are accurately reflected in the unit test cases.

Code reviews and reviews of the unit test cases are suitable means for consistency checking. Establishing vertical, bidirectional traceability makes consistency checking possible, since the design specifications valid for a particular software unit are completely known. Traceability must of course also be maintained accordingly during development (for traceability see also section 2.24).

BP7: Ensure consistency and bilateral traceability of software requirements to software units. Ensure consistency of software requirements including verification criteria to software units including verification criteria. Consistency is supported by establishing and maintaining bilateral traceability between the software requirements including verification criteria and software units including verification criteria.

NOTE: *Consistency and bilateral traceability need only be established between software requirements and software units for requirements that cannot be addressed in software detailed design (e.g., non-functional requirements, attributes, etc.).*

BP7 is only relevant if the call for consistency and traceability is not already covered by BP6, or if a requirement is not reflected in the software detailed design (see note). In this case, consistency of each completed software unit with the software requirements must be ascertained. Consistency checks must ensure the following:

- All relevant requirements appertaining to a particular software unit have been taken into account in the unit-test test cases and
- For each requirement, the associated verification criteria (see ENG.4 BP2) have been considered during test case design.

An appropriate means for consistency checking is the review of the unit test cases. Establishing bidirectional traceability facilitates consistency checking, since the requirements associated with a particular software unit are completely known. The existence of—in this case vertical—traceability has additional advantages:

- Requirements and associated verification criteria can already be consulted for background information during coding, or no requirement is missed.
- Verification criteria can already be used during the creation of unit test cases.
- Checking if all requirements are covered by unit tests is made easier.
- Identification of affected software units in case of requirements changes is easier.

Of course, traceability must be maintained during all development phases (see also section 2.24 for more details on traceability).

BP8: Ensure consistency and bilateral traceability of software units to test specification for software units. *Ensure consistency of software units including verification criteria to test specification for software units including test cases for software units. Consistency is supported by establishing and maintaining bilateral traceability between the software units including verification criteria and test specification for software units including test cases for software units.*

Typically, a whole range of tests exists for a software unit. This base practice is supposed to ensure that the verification criteria of a software unit are consistent with its test specification. The term »test specification«[41] (see work product 08-50) denotes a whole range of contents including, in brief, the concept of how to test a software unit, how to execute test cases, as well as individual test cases including regression test cases.

Establishing horizontal, bidirectional traceability makes consistency checking possible, since relevant tests and test cases associated with a particular software unit are completely known. If a software unit includes several functions, traceability becomes two-tier (software unit–function–test cases). Of course, traceability must be maintained accordingly throughout the development phases (for traceability see section 2.24).

Another important means for consistency checking is code coverage analysis (see glossary), which checks if test coverage goals have been reached.

---

41. The existence of a test specification is indirectly required by BP8, its creation is not described in ENG.6.

2.8.4 Designated Work Products

08-50 Test Specification
See ENG.8

08-52 Test Plan
See ENG.8

11-05 Software Unit
A software unit is a piece of code containing operations and data for the implementation of a more or less self-contained task. A software unit depends on software design, programming language, and application. Communication with the outside world is only possible via defined and explicitly specified interfaces. In Automotive SPICE, a software unit is the smallest software building block.

13-22 Traceability Record
See ENG.2

13-50 Test Result
See ENG.8

2.8.5 Characteristics of Level 2

On Performance Management
Individual software units are developed on the basis of the software design. This development usually consists of many small, iterative work steps so that detailed process planning makes little sense. This is why often only start and completion are planned per individual software unit. If a developer has to create several smaller software units (e.g., taking only a few days for each unit), even more condensed planning is sufficient. Another part of performance management is planning and tracking of the used resources and schedules.

On Work Product Management
The requirements of Process Attribute PA 2.2 particularly apply to the main work product of the process: the code. Its verification is already effected by the base practices. Requirements on further work products, such as verification procedures, test cases and test results, can be covered by defined development processes and document templates. They can be checked in reviews. Work products, particularly code and associated documentation, are put under configuration control. For each baseline, all associated work products are specified.

**Excursus: Test Documentation According to IEEE-Standard 829-1998
(Software Test Documentation)**

The IEEE Standard 829 has become established as an industry standard, and the terms used there are commonly used in the industry. Automotive SPICE, too, uses the IEEE definition for some of the work products. The following list provides a brief overview of the terms used in the IEEE 829 standard; references in brackets are to the WP-ID.

- Test Plan (Test Plan, WP-ID 08-52): Planning of all test activities. Among other things, the test plan identifies the components and functions to be tested, functions not to be tested, method, pass or fail criteria, products, test activities, test environment, responsibilities, staff, and schedule.
- Test Design Specification (Test Design Specification, WP-ID 08-50): Concept of how components are to be tested. The test specification includes, among others, features to be tested, specific test techniques, pass or fail criteria, and a list of associated test cases.
- Test Procedure Specification (Test Procedure Specification, WP ID-08-50): Specification of the steps necessary for performing a test. Among others, the test procedure specification includes the purpose and a list of all associated test cases, preconditions and preparatory steps, as well as guidelines for executing the procedure and logging of the results.
- Test Case Specification (Test Case Specification, WP-ID 08-50): Description of how a particular test object (e.g., a function) is to be tested. Among others, the test case includes the test items, input specifications, output specifications, and environmental needs.
- Test Item Transmittal Report[42]: Description of the delivered software (e.g., to the test team or customer). It includes, among others, the exact content of the delivery with versions of the individual items, associated documentation, persons responsible for each item, method of transmittal of the software, changes against the previous delivery, known defects, and approval.
- Test Log (Test Log, part of WP-ID 13-50): Includes the list of tested items and version, test environment, list of test cases (tester, date, results, possibly unexpected events).
- Test Incident Report (Test Incident Report, part of WP-ID 13-50): Description of problems[43] detected during testing; The test incident report contains the same data as the test log but is more detailed to support root cause analysis. It

---

42. Also known as »Release Notes«. Here Automotive SPICE does not adopt the IEEE definition but provides its own definitions in SPL.2 BP12 and WP-ID 11-03.
43. A problem need not necessarily originate from a software defect. Causes may be: developers and testers interpret specifications differently, developers and testers use different system and test environments, fault in the test case.

is further processed, applying the problem resolution management process (SUP.9).
- Test Summary Report (Test Summary Report, part of WP-ID 13-50): Provides project personnel and management with an overview of the test results; including, among others, identification of the test items and how they were tested, a summary of the results and their evaluation (passed/not passed, failure risks, etc.).

2.9 ENG.7 Software Integration Test

2.9.1 Purpose

The purpose of the software integration test process is to integrate the software units into larger assemblies, producing integrated software consistent with the software design and to test the interaction between the software items.

Software integration should be done in steps and be accompanied by tests. Tests complement the preceding unit tests. Stepwise integration helps, particularly in complex systems, to identify and remove defects as early as possible. This way the »progress of software integration« is secured and the functional growth remains manageable during integration. Regarding documentation, Automotive SPICE follows the IEEE-terminology (see also our excursus »Test Documentation« in ENG.6.)

2.9.2 Characteristics Particular to the Automotive Industry

In Automotive SPICE, aspects of integration are described in ENG.7 and also in ENG.9 (System integration). If an integrated system consisting of hardware and software items is developed in the project, pure software integration is described in ENG.7, and software/hardware integration in ENG.9. In some cases, where hardware is already available as a finished product (for instance, as an ECU), integration only takes place once, and the different software units are directly integrated onto the target hardware.

> **Note to Assessors**
>
> If ENG.7 and ENG.9 are merged, integration activities can be rated either in ENG.7 or ENG.9. To assess both (identically) does not make any sense. In most cases rating will be done in ENG.9, and ENG.7 will be declared non-applicable. The different aspects of the respective integrations levels should be kept in mind, though. For instance, interfaces between hardware and software are primarily considered during system integration. Interfaces between software items are analyzed during software integration.

Many organizations work with a high software reuse rate, basing the project on an existing software platform that is then modified or enhanced to meet project requirements. Compared to a purely project-specific development, increased obligation of care applies with respect to quality assurance issues while developing the platform software further. This obligation applies to the ENG.7 process and also for all other engineering processes. If a modeling tool is used for a model-based software design, software integration can be performed directly in the modeling tool itself. Integration tests performed in the modeling tool are generally known as software-in-the-loop tests (SIL[44] tests).

2.9.3 Base Practices

BP1: Develop software integration strategy. Develop the strategy for integrating software items consistent with the release strategy and an order for integrating them.

This strategy specifies the stepwise sequence in which the software units are integrated into increasingly larger, integrated software items, taking into account the release strategy (see SPL.2[45]). From a certain aggregation level onwards, the integrated software items correspond to elements of the software architecture. The integration strategy must therefore be compatible with, or be derived from, the software design and the software architecture. Depending on the product, different integration strategies are possible, whereby all strategies should consider different approaches for the integration of new or changed software items:

- Bottom-up, starting with hardware-related software (see also figure 2–9)
- Top-down, starting with the user interface
- Starting with a basic software, then integration of critical modules/units
- Integration in any sequence, e.g., according to availablity
- Integration of all parts in one single step

---

[44]. Not to be confused with SIL = Safety Integrity Level (see chapter 5).
[45]. The term »release strategy« is used in Automotive SPICE, but not defined (see »release planning« instead).

2.9 ENG.7 Software Integration Test

The last method is also called, somewhat derogatively, the »Big Bang« method. It only makes sense in flat architectures where only a few coexisting and functionally independent modules are integrated with a software base. This strategy is questionable in large, complex systems and/or in distributed development with different development teams.

In contrast, a specific integration sequence does make sense if one needs to follow the »onion layer principle« to proceed from the inside out, because functions in a layer can only reasonably be tested if the functions of the layers below already function reliably.

> **Note to Assessors**
>
> If software integration is implemented in a project using the »Big Bang« strategy, the project must prove beyond doubt the appropriateness of this method. Reasons like cost or time pressure alone are not proof enough. If the team of assessors comes to the conclusion that »Big Bang« is not the appropriate method or that it may have a negative impact on product quality, the corresponding practices should be down-graded and the reasons for this decision be documented.

Furthermore, the sequence is specified in which the software items are to be integrated. The integration strategy must therefore also be compatible with priorities regarding the software requirements, e.g., with a release strategy that prescribes which requirements are to be implemented in which releases. The integration sequence may have to be adapted depending on the functional increase and possible structural changes in the course of the project.

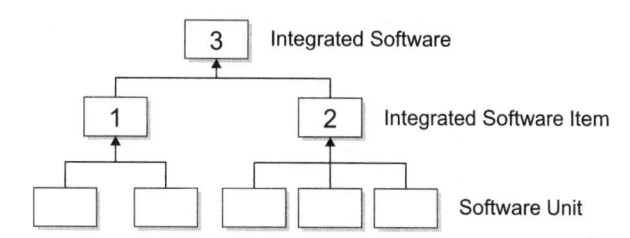

Figure 2-9   Integration sequence following a bottom-up strategy

The integration strategy can be defined in different ways. In some cases it is not required in written form, e.g., in cases where the »Big Bang« approach is applied or where there are only a few levels and where because of the architecture there is no doubt as to which module belongs to which level, and if the team is very familiar with the integration sequence.[46]

---

46. In automotive OEM assessments, documented evidence is expected in most cases. At Level 2, at least scheduling is expected to be documented.

BP2: Develop software integration test strategy. *Develop the strategy for testing the integrated software items. Identify test steps according to the order of integration defined in the integration strategy.*

NOTE: *The software integration test will focus mainly on interfaces, data flow, functionality of the items, etc.*

NOTE: *The software integration test process should start with the beginning of the software development process. There is a close link from Software Requirements Analysis ENG.4, Software Design ENG.5 or Requirements Elicitation ENG.1 in developing test cases and testable requirements.*

NOTE: *The Software integration strategy contains different approaches of integrating software items depending on the changes (e.g., new units, changed units). The integration strategy also includes the most suitable test methods to be used for each integration approach.*

NOTE: *The identified items and the order of integration will have an influence on integration test strategy.*

Integration testing focuses on the interfaces between the software items, the data flow in-between, and (if not covered by ENG.6) cross-functionality. Integration testing is primarily performed against the specifications of the software architecture. The functionality of the individual software units is already verified during unit testing (ENG.6); the functionality of the integrated software is verified during software testing (ENG.8).

Integration is naturally carried out in every project; however, integration testing as a separate test level is sometimes missing. This may certainly make sense in simple software architectures where test levels may have been combined (on this point, see ENG.6 BP1), and if integration testing is explicitly included there.

> **Note to Assessors**
>
> The existence of a separate test level for integration tests is not mandatory; however, the actual purpose of integration testing (see above) must be satisfied. This can be done with an appropriately designed software test covering integration test aspects (see also our deliberations shown in ENG.6 BP1).

Already at the beginning of the software development process, during requirements elicitation (ENG.1), software requirements analysis (ENG.4), and software design (ENG.5), attention should be paid to the integration test strategy. This way, appropriate verification criteria can be defined for integration testing at an early stage. The nature of the modifications on individual software items influences the nature of integration testing—in case of new software items there must be exhaustive integration testing. In case of changes, only those parts need to be tested that were also specified in the regression test strategy (see BP8). The available software items and the integration sequence also influence the integration test strategy in the way that the integration sequence largely reflects the test sequence. In some cases, independent software items can already be pre-integrated and tested prior to integration. However, the natural integration test sequence corresponds with the integration sequence.

> **Note to Assessors**
>
> It is advisable to coordinate and jointly document the test strategies of the test activities described in ENG.6, ENG.7, ENG.8, ENG.9 and ENG.10 within an overall test strategy. It is not necessary to present a separate strategy document in each case. However, if an overall test strategy exists, all aspects of the relevant sub-processes including regression must be considered.

Automotive SPICE describes the test strategy in the context of the test plan. The test plan (see our excursus »Test Documentation« in ENG.6) should include the following items:

- What needs to be done in what way? This includes planning of test activities, test strategies, test methods, test sequences, regression tests, components and functions to be tested, functions not to be tested, and integration into the software development process.
- Which quality standard is the project to achieve or to observe (e.g., definition of test completion criteria)?
- How much time and staff are needed or available?
- What are the means available for testing?
- Which risks and which mitigating measures are there?

BP3: Develop test specification for software integration test. Develop the test specification for software integration test including test cases, to be executed on each integrated software item. The test cases should demonstrate compliance to the software architectural design and software detailed design allocated to each software item.

For each integrated software unit the tests to be performed and the method must be specified.[47] This includes a description of the tests written in a way that the tester knows exactly how tests are to be executed, also the environmental set-up, operational procedures, input data or data to be used, as well as a description of expected results or behavior. Here, preliminary work like the verification criteria out of the ENG.5 process, can be used and worked out in more detail. Integration testing focuses on testing the interaction of the software units/items. Testing is

---

47. Contentwise, this corresponds to the IEEE-terms test design specification, test procedure specification, and test case specification (see excursus »Test Documentation« below ENG.6).

done primarily on cross-module functions, interfaces, data flows, etc., to prove that the design requirements have been met. These tests typically take place in a laboratory environment. Since they require good knowledge of the internal software structure, so-called white-box tests[48] and gray-box tests are typically used.

In most cases, testing is performed by developers of the respective software units/items after integration. From a certain project size onwards, especially if development is distributed across different locations, we find the role of an »integrator« or of an »integration representative« to be a beneficial addition. In this case, additional test staff for an integration test group will frequently be available.

Examples of code verification criteria are:
- Fulfillment of the software requirements
- Compliance with coding guidelines
- Consistency with software design
- Consistency of the external interfaces
- Consistency internally between software units/items
- Fulfillment of criteria regarding the degree of test coverage

BP4: *Integrate software units and software items.* *Integrate the software units to software items and software items to integrated software according to the software integration strategy.*

NOTE: *Software units are integrated to software components[49] and software components to integrated software.*

NOTE: *The integration of the software units and software components also integrates their data. Data can be calibration data and variant coding data.*

Integration is performed according to the integration strategy and the project plan. Integration of the associated calibration data and variant coding data[50] plays an important role in the automotive setting, since the functionality of the software is highly dependent on this data. Calibration data is developed or matured during project-specific vehicle testing; in some projects it is created wholly or in part by the OEM.

> **Note to Assessors**
> The integration of data/software is an integral part of the integration process. It must therefore be examined and is rating-relevant.

---

48. See excursus »Test Methods« below ENG.8.
49. The term »software components« is an inconsistency in the original text and used synonymously for »software items«.
50. This refers to data which affects the execution of the software. Calibration data are used to fine-tune the behavior of cars (e.g., the way the car starts to move, or the manner of ESP intervention) and are determined during test drives. Variant coding data allow the use of the same software for different body types, country variants, right/left-hand drives, etc.

BP5: Verify the integrated software. Verify each integrated software item against the test cases for software integration test according to the software integration test strategy.

NOTE: *Verification of the integrated software produces the test logs.*

Tests are performed according to the integration test strategy, and the results are documented in test logs. A positive test result is achieved if accomplished and expected (planned) results are identical. If there are deviations from expected results or if results are completely different, the associated test cases are considered negative or failed. Regarding the resulting test logs (part of WP-ID 13-50), see excursus »Test Documentation«, ENG.6. As a minimum, the following information is contained in a good test log:

- Who tested which software version, when, and in which test environment?
- Which test cases were executed in which sequence and what were the results (e.g., positive, negative or unexpected)?
- If defects were detected, i.e., if tests failed: Creation of a test incident report stating the defect ID; this report is used for problem management tracking purposes (see BP6).

BP6: Record the results of software integration testing. Document the results of software integration testing and communicate to all relevant parties.

NOTE: *The test incident reports and the test summary report are based on the test logs.*

Based on the test logs referred to in BP5, test incident reports (part of WP-ID 13-50) are created for failed test cases. A test summary report (part of WP-ID 13-50) is created for the overall test.

A test incident report need not necessarily be based on a defect in the tested system. The discrepancy between actual and expected results may have a variety of different causes. Expected values could be wrong, a test case may have been wrongly executed, or requirements may have been misunderstood because they left too much room for interpretation. A test incident report comprises at least the following information (see also SUP.9):

- All available details, i.e., the underlying test log including expected and actual values
- Defect-ID
- If possible, an analysis of the problem's impact on other test cases
- Additional information that helps in carrying out root cause analysis

There is no 1:1 relationship between test logs and test incident reports. A negative test may have several causes and result in several test incident reports. On the other hand, several negative tests may be due to one cause. In this case it is important to create separate test incident reports according to the impacted functions, since in most projects the individual functions are handled by different people. This is the only way in which the defects in question can be consistently tracked and removed. All relevant information about the software integration test is summarized in the test summary report. A good test summary report comprises at least the following information:[51]

- Date, duration, scope or effort of the software integration test
- Number of passed and failed test cases, including the number and priority of problem cases identified during testing
- Number of planned and executed tests (e.g., number of iterations and necessary regression tests)
- Evaluation of how well software integration testing was performed and what the quality of the software is
- Summary and evaluation of the results and whether the test as a whole is considered passed or not

Based on the test summary report, a decision is made in the end whether the quality of the integrated software is considered sufficient enough to proceed with the next process step ENG.8 (Software test).

Records of test results are often also required by the customer because they are taken into account during defect tracking or used in joint project meetings. Furthermore, test records are performance records used for progress control and are usually required as evidence during the development of safety related systems. More importantly, they form the basis for oncoming repair work and revisions.

BP7: *Ensure consistency and bilateral traceability of software architectural design and software detailed design to software integration test specification.* *Ensure consistency of software architectural design and software detailed design to software integration test specification including test cases. Consistency is supported by establishing and maintaining bilateral traceability between software architectural design and software detailed design to software integration test specification including test cases.*

Consistency between software integration tests and software design is primarily guaranteed by »horizontal« traceability (see also section 2.24) between the software architecture and the test specification, or the test cases of the software integration test, respectively. This requires specification of which architecture element was tested with which test cases of the software integration test, and vice versa. Traceability must be maintained throughout the life of the project and particularly after changes.

---

51. See excursus »Test Documentation« in ENG.6.

2.9 ENG.7 Software Integration Test

BP8: Develop regression testing strategy and perform regression testing. Develop the strategy for re-testing the software items if changed software items are integrated. Perform regression testing as defined in the regression test strategy and document the results.

Regression tests are necessary to ensure that attributes already secured by previous testing are still available after code changes (e.g., as a result of defect removal or function enhancements).[52]

A regression test strategy, in its most trivial form, consists of maintaining a collection of test cases which are executed by default after code changes[53], in addition to newly developed tests particularly aimed at covering modifications. In most cases, such test case collections are very comprehensive, and a lot of effort needs to be expended for their execution if tests are not automated. In iterative development models, changes and especially functional enhancements belong to normal life; this is why more intelligent strategies are necessary, e.g.,:

- Only defined subsets of the collection of test cases are tested, depending on *where* the change/enhancement was made.
- Only defined subsets of the collection of test cases are tested, depending on *the type of delivery.*

> **Note to Assessors**
>
> In an assessment it is important for rating purposes if regression testing is carried out to the required extent and not how intelligent or labor-saving the underlying strategy is.

As a rule, the regression test strategy should be documented in the test plan, even in trivial cases where, for instance, a renewed 100% test is repeated.

Since the regression test strategy is required in different processes (ENG.7 to ENG.10), it makes sense to establish a common regression test strategy for all engineering processes. Execution and results of regression tests are of course comprehensibly documented even if this is not explicitly mentioned in Automotive SPICE.

Case Study: Software Integration at XY Ltd.

In order to better understand the interrelationships of this process, we have described a typical procedure for software integration, integration test and regression test in the form of a case study:

---

52. In other words: Does what verifiably worked before the change still work now?
53. In other words: All previous tests are executed again.

XY Ltd. pursues a two-tier integration strategy: At each development site (tier 1) and at the development headquarters (tier 2), the integration sequence is specified and documented in writing. Moreover, for each of the tests to be executed at the respective integration level, there is a description of the tests to be executed in form of a so-called »test spec«. It includes the respective test method and a collection of test cases. The test spec has a column for the description of the test results that needs to be filled in manually during test execution. It is subsequently scanned in and stored in the configuration management system. Additionally, a »test summary document« is drawn up, including a quantitative evaluation of the results (number of tests, number of passed/not passed test cases, brief verbal summary); this is also filed in the CM system. Design documents reference all code files by name, and code files reference all design elements. Another column in the test spec is for regression tests to specify the affiliation of the test case to either a short test or a long test. The regression test strategy (central document to be applied in all locations) prescribes a short test for weekly releases and a long test for monthly releases and customer deliveries. The test specs are constantly updated to cover enhanced or new functionality.

At each location and at headquarters there is an integration manager. For each site, integration period and staff effort are planned in detail. All test cases of the test spec must be successfully passed. Dates for integration level test execution at the different locations and due-dates for delivery to headquarters are specified in a joint MS-Project plan and monitored by the local or central integration manager. Furthermore, there are daily teleconferences. Responsible are the integration manager and other staff. All responsibilities are documented in the MS-Project plan. External companies are controlled at headquarters by dedicated members of staff (»subcontract managers«) who keep in close contact with their respective companies. Their subcontracted deliveries are implemented in the releases only at defined milestones. In all other respects, subcontract managers are treated like location integration managers.

2.9.4 Designated Work Products

01-03 Software Item

A software item consists of integrated software units, configuration files, data and associated documentation (see figure 2–2 and related comments).

01-50 Integrated Software

Integrated software is understood to be an aggregation of software items, the set of executable files for a particular ECU configuration, and possible associated documentation and data (see figure 2–2 and related comments).

08-50 Test Specification
See ENG.8

08-52 Test Plan
See ENG.8

13-22 Traceability Record
See ENG.2

13-50 Test Result
See ENG.8

17-02 Build List
The primary purpose of the build list is the identification of software aggregates and required system elements (parameter settings, macro libraries, data bases, job control languages, identified input and output source libraries, etc.). It also documents the sequence of actions or activities required for the software build.

2.9.5 Characteristics of Level 2

On Performance Management

Software integration planning and tracking is done using the project management methods applied in the project. In small projects, at least resource and schedule planning should be performed. An additional requirement in large projects is that the individual integration steps are planned in detail.

On Work Product Management

The requirements of Process Attribute PA 2.2 particularly apply to the test plans, the test specifications, and the test results. Baselines are drawn according to plan.

2.10 ENG.8 Software Test

2.10.1 Purpose

The purpose of the Software testing process is to confirm that the integrated software meets the defined software requirements.

Tests ensure that the integrated software corresponds with the software requirements identified in ENG.4. This refers to the integrated software product after integration (ENG.7) has been completed. It is quite possible that there are several parallel software products running on different processors, and that these are integrated with other system components (i.e., including hardware, mechanics) during later system integration (ENG.9).

2.10.2 Characteristics Particular to the Automotive Industry

It is common practice in the automotive industry to develop integrated systems consisting of hardware, software, and mechanics components.[54] In an assessment, the question arises how to relate the different tests performed in the project to the respective processes (ENG.7, ENG.8, ENG.9, ENG.10). This is easily done if there are explicit software requirements with clearly associated tests. In reality, things are often different: requirements describe functionality realized by the interaction of hardware and software, which is then later tested after integration.

> **Note to Assessors**
>
> It is common assessment practice to assign tests as follows:
> - ENG.7: Tests during software integration
> - ENG.8: Tests of finished software products prior to integration with hardware, whereby the target hardware is emulated by software and/or hardware lab set-up. The lab set-up may also comprise the finished ECU.
> - ENG.9: Tests during system integration of software and hardware components
> - ENG.10: Tests of the integrated system on the target hardware
>
> We frequently encounter the following scenario: The target hardware is already available and all tests are directly executed on the target hardware. In this case, ENG.7/ENG.8 and ENG.9/ENG.10 are mostly redundant. Assessments will then usually waive double evaluation and assess integration and test only in ENG.9/ENG.10, declaring ENG.7/ENG.8 to be »not applicable«. Attention should be paid to the different aspects of the respective integration level: Test cases that are based on ENG.4/ENG.5 must be executed in ENG.9/ENG.10, i.e., conformity with software requirements and software architecture must now be evidenced at system level. If this is not done it will have negative consequences when evaluating ENG.9/ENG.10.

If a prototype is delivered (e.g., an early sample) which does not contain the complete functionality, the software tests must ensure that the delivered (partial) functionality works according to the requirements. It is a particular characteristic in automotive software development that hardware-relatedness and real-time behavior (see also ENG.5, section 2.7.2 »Characteristics Particular to the Automotive Industry«) have a strong impact on the type of test and the test environment (often the target hardware and target environment).

---

54. There are, however, projects that deliver a software-only product to be integrated by another company (e.g., at the customer's) in a complete system composed of hardware, software and mechanics.

2.10.3 Base Practices

BP1: Develop software test strategy. Develop the strategy for software testing consistent with the release strategy.

Prior to testing the entire software, the targets and general parameters applicable to software testing must first be documented in a test plan[55] (see excursus »Test Documentation« in ENG.6). The release strategy defines at what time which functionality is going to be available, hence, at what time which tests can and must be executed.

BP2: Develop test specification for software test. Develop the test specification for software test including test cases, to be executed on the integrated software. The test cases should demonstrate compliance to the software requirements.

NOTE: *The Software testing process should start early in the software development life cycle. There is a close link to Software Requirements Analysis ENG.4, Software Design ENG.5 and Requirements Elicitation ENG.1 in developing test cases and testable requirements.*

The tests to be executed must be specified. This corresponds with the information content of the test design specification, test procedure specification, and test case specification (elements of the test specification, WP-ID 08-50) documents. During specification, parts of the test plans from BP1 may be supplemented. In total, the following is required:

- How is testing done? (methods, groups of tests, guidelines for test execution, test sequence, test end criteria, etc.)
- Which environment/conditions, actions, and input data are used by the test?
- Which requirements are verified by the different tests?
- Which behavior (e.g., in the form of output data) is expected from the integrated software product to be able to declare that a test was successful?

All in all, the tests as a whole must be suitable to prove that all software requirements have been implemented. From a methodological point of view, white-box as well as black-box testing could be used here. In practice, functional tests, i.e., black-box tests, are performed in most cases. Preparations for the software test process and the test strategy should already start at the beginning of the development process and be further refined as a result of a growing understanding resulting from requirements analysis. This way, appropriate verification criteria can be defined for software testing at an early stage in development.[56]

---

55. Different elements of the test plans according to IEEE are not addressed by Automotive SPICE until Level 2 with the generic practices of PA 2.1.
56. Early derivation of verification criteria from requirements may be regarded as requirements quality assurance. This way, inconsistencies and room for interpretation are frequently found at an early stage.

BP3: Verify integrated software. *Verify the integrated software against the test cases for software testing and according to the software test strategy.*

NOTE: *Verification of the integrated software produces the test logs.*

NOTE: *Tests should be automated as far as possible having regard to efficiency.*

Planned tests are executed and test logs are created. See also our elaborations related to the documentation of test results given for ENG.7 BP5, and the excursus »Test Documentation« in ENG.6.

BP4: Record the results of software testing. *Document the results of software testing and communicate to all relevant parties.*

NOTE: *The test incident reports and the test summary report are based on the test logs.*

Software test results are documented in the test logs and test incident reports, and then incorporated in a test summary report (in line with ENG.7 BP6). Based on the test summary report, a decision can be made as to whether the software quality is sufficient enough to continue with the system integration (ENG.9) process step. It may become apparent during bug fixing that the underlying software test specifications (BP2) and perhaps the requirement specifications, too, require revision.

BP5: Ensure consistency and bilateral traceability of software requirements to software test specification. *Ensure consistency of software requirements to software test specification including test cases. Consistency is supported by establishing and maintaining bilateral traceability between the software requirements and software test specification including test cases.*

NOTE: *Consistency can be demonstrated by review records.*

Consistency between software requirements and the software test specification is assured by means of »horizontal« traceability (see also section 2.24), meaning that for each software requirement a corresponding software test must be specified which covers that particular requirement. Equally, for each software test case specification it must be clear which requirements it is going to test. Proof of such coverage can be accomplished through reviews and corresponding review records. Here, too, traceability must be retained during the whole project lifecycle—especially if changes have been made.

BP6: Develop regression test strategy and perform regression testing. *Develop the strategy for retesting the integrated software should a software item be changed. If changes are made to software items carry out regression testing as defined in the software regression test strategy, and record the results.*

This practice requires regression testing for the complete, integrated software based on changes made to software items. On the principal subject-matter of regression testing, see ENG.7 BP8.

Excursus: A Brief Overview of Test Methods

- Static analysis: analysis of the code with the help of the development tool suite or by means of other specialized tools, for instance, to detect »dead« pieces of code, infinite loops, non-initialized variables, violations of coding conventions, and so on. This analysis is called »static« because code is not being executed during the analysis.
- Dynamic method: In dynamic testing, code is executed in a test environment. The following types of tests are distinguished:
  - White-box tests (also called »structure-based tests«) are derived from knowledge of the internal structure of the software based on the program code, design, interface descriptions, etc. In most cases they are performed by the software developers themselves as they know the internal structure of the software components very well. One risk is the »organizational blindness« of the software developer.
  - Black-box tests (also called »functional tests«) compare the externally observable behavior at the external software interfaces (without knowledge of the structure) with the desired behavior. Black-box tests can also be performed by a separate test team or by an external tester. One advantage is that the »four-eyes-principle« is being observed.
  - Gray-box tests (a combination of black/white-box test) are typically provided by the software developer. Like black-box tests, they aim at the externally observable behavior of the software. Knowledge of the inner structure of the software unit helps to design these tests.

Excursus: Some Methods for the Derivation of Test Cases

- Based on the processing logic: The processing logic refers to the consequences of actions and decisions (so-called paths), whereby decisions are made based on conditions. Different test coverage grades can be distinguished:
  - Statement Coverage (C0 metric): Each action (= statement) is executed at least once.
  - Branch coverage (C1 metric): Each action is executed at least once, and each possible decision is induced at least once.
  - Condition coverage: Each action is performed at least once, and each possible result of a condition is induced at least once.
- Equivalence classes: Differentiation of input values into classes within which each value is equally likely to detect a defect.
- Boundary value analysis: Empirically speaking, errors are more likely to occur at boundaries. This can, for instance, be tested by a value directly on, above, or below the limit.

2.10.4 Designated Work Products

Summary of Test-Relevant Work Products

For clarity's sake, we shall summarize here the work products relevant for testing. There are three central work products for the test processes in Automotive SPICE: the test specification, the test plan, and the test results. These work products can for the most part be attributed to terms defined by the IEEE (see excursus »Test Documentation« in ENG.6):

08-50 Test Specification

The Automotive SPICE test specification includes the following:

- IEEE Test Design Specification
- IEEE Test Procedure Specification
- IEEE Test Case Specification
- Labeling of regression test cases according to IEEE

and additionally for integration tests:

- Identification of system elements to be integrated
- Integration sequence

08-52 Test Plan

The Automotive SPICE test plan comprises the following components:

- IEEE Test Plan
- Test Strategy, e.g., black-/white-box tests, equivalence class tests, regression test strategies

13-50 Test Result

The Automotive SPICE test result contains the following:

- IEEE Test Logs
- IEEE Test Incident Report
- IEEE Test Summary Report

2.10.5 Characteristics of Level 2

On Performance Management

The conceptional part of test planning (i.e., what is tested and how) is covered by BP1; scheduling and tracking of individual test activities is required by the generic practices of Level 2, and the project management process. In small projects, test planning is usually part of project management. As most larger projects have their own test subproject, detailed planning and tracking is typically done by subproject management, while project management does rough planning only.

On Work Product Management

The requirements of Process Attribute PA 2.2 particularly apply to the test plan, the test specification, and the test results. Requirements concerning work products (i.e., test plans) and quality criteria (e.g., for reviews) are defined, relevant documents like test cases are under configuration control, reviews are conducted and can be verified. Modifications of work products are managed in a verifiable way, including, for instance, the controlled change/endorsement of test plans and test cases to safeguard changes made to the software.

2.11 ENG.9 System Integration Test

2.11.1 Purpose

The purpose of the System integration test process is to integrate the system elements to produce an integrated system that will satisfy the system architectural design and the customers' expectations expressed in the system requirements.

At this point in development, the system elements, i.e., software, hardware, and mechanics, are integrated to provide the functionality required of the system. It now becomes apparent whether these components match. With the exception of a few special features, the system integration test process ENG.9 is methodologically identical to the software integration test process ENG.7, yet it relates to different objects. The following deliberations focus primarily on the differences to ENG.7.

2.11.2 Characteristics Particular to the Automotive Industry

In the simplest case (one software, hardware, mechanics), integration of hardware, software and mechanics to one system is accomplished in one single step through mechanical integration and software flashing, followed by testing. In case of complex products (multiprocessor architectures, several hardware and mechanics components, several software components), this integration is usually done in several steps.

The definition and conceptual demarcation of the term 'system' is rather difficult (see also section 2.4.2), since it depends very much on the areas where the processes are applied. At the top level (OEM), the system constitutes the vehicle plus perhaps some additional external systems (e.g., telematics). Regarding an individual vehicle component, for instance, an ECU with peripheral devices, the term 'system' comprises the component's individual elements plus its interfaces to other vehicle components. As far as the mechanics of the ECUs are concerned, its interfaces consist of fixtures and mounting space. Concerning hardware, interfaces consist of cables and connectors, whereas software interfaces consist of the communication protocols. In practice, two factors are particularly problematic:

- Responsibility for integration is not accurately defined or agreed between OEM/supplier or tier-one-/tier-two-supplier.
- Due to simultaneous development, finished versions of third party components are usually not available for integration since they are also still under development. Moreover, in test vehicles we find that in most cases configurations of involved parties (OEM, suppliers) vary because of different hardware and software versions. The consequence is that failures occur with party A and not with party B, and vice versa.

In iterative development, the number of implemented functions increases with each delivery. However, weeks or months may pass between delivery dates. The project must support this mode of practice, for instance, with suitable release planning of hardware, mechanics, and software. This is to ensure that functional interim versions are delivered in line with agreements, and that the customer can conduct his own sample tests. The project can thus appreciably increase the quality and useability of the fully developed overall system (see also our deliberations in sections 2.9.2 and 2.10.2 »Characteristics Particular to the Automotive Industry« in the ENG.7 and ENG.8 processes).

2.11.3 Base Practices

BP1: Develop system integration strategy. Develop the strategy for integrating the hardware items and integrated software consistent with the release strategy and an order for integrating them.

Our deliberations shown in ENG.7 BP1 equally apply to this base practice. Now the objects to be integrated or tested are both software and hardware elements[57] and their different integration levels. In the simplest case, software and hardware are developed simultaneously and are subsequently integrated. In complex cases, the complete system consists of an aggregation of several electromechanical and electronic elements (the latter with software items), so that several separate hardware integrations, and hardware and software integrations take place[58]. Particular demands must be placed on these strategies if the system elements come from different development partners. Corresponding requirements in ENG.7 and ENG.9 can of course be implemented in a joint integration strategy.

---

57. It is true that mechanics was not mentioned here; however, it must likewise be integrated.
58. Software-only integration is described in ENG.7.

BP2: Develop system integration test strategy. Develop the strategy for testing the integrated system. Identify test steps according to the order of integration defined in the integration strategy.

NOTE: The integration test will focus mainly on interfaces, data flow, functionality of the system elements, etc.

NOTE: The System integration test process should start with the beginning of the system development process. There is a close link from System Requirements Analysis ENG.2, System Architectural Design ENG.3, or Requirements Elicitation ENG.1 in developing test cases and testable requirements.

NOTE: The system integration strategy contains different approaches of integrating system elements depending of the changes (e.g., changed hardware items, new integrated software). The integration strategy also includes the most suitable test methods to be used for each integration approach.

NOTE: The identified system elements and the order of integration will have an influence on system integration test strategy.

The deliberations shown in ENG.7 BP2 also apply to this base practice. Now the elements under test are hardware units and integrated software instead of integrated software items only. Similarly, the system integration test strategy should already start during system requirements analysis/architecture design. Here, too, the test strategy is documented as part of the test plan—an overall strategy covering all tests is advisable.

BP3: Develop a test specification for system integration. Develop the test specification system integration, including the test cases, to be executed on each integrated system element. The test cases should demonstrate compliance to the system architectural design.

Our deliberations concerning software shown in ENG.7 BP3 are also valid here. It should also be specified which other system elements must be available in hardware and/or software to be able to test a specific system element.

BP4: Integrate system elements. Integrate the system elements to an integrated system according to the system integration strategy.

NOTE: The system integration can be performed step wise integrating the hardware elements as prototype hardware, peripherals (sensors and actuators) and integrated software to produce a system consistent with the priorities and categorization of the system requirements.

Integration is performed in steps at system level according to the system integration strategy, the project plan, and according to the release plan (see SPL.2).

*BP5: **Verify the integrated system**. Verify each integrated system element against the test cases for system integration according to the system integration test strategy. Demonstrate that a complete set of useable deliverable system elements exists, is constructed.*[59]

NOTE: Verification of the integrated system produces the test logs.

NOTE: Verification of the integrated system can be performed using environment simulation methods (e.g., Hardware-in-the-Loop-Simulation, vehicle network simulations).

Tests are carried out in line with the integration and regression test strategy, and the results are documented (see also ENG.7 BP5). Standard practices used in the automotive industry for system integration testing deploy simulations like the partial assembly of vehicle components on one or several Hardware-in-the-Loop test benches (HIL), or rest bus simulations where software simulates unavailable components. Apart from this, system integration testing in test vehicles is also possible, focusing, for instance, on interfaces.

*BP6: **Record the results of system integration testing**. Document the results of system integration testing and communicate to all relevant parties.*

NOTE: The test incident reports and the test summary report are based on the test logs.

Our deliberations shown in ENG.7 BP6 apply here, too.

> **Note to Assessors**
>
> Of course, the relevant parties to whom the results must be communicated are different from those involved in software integration testing (ENG.7). This must be borne in mind, particularly if ENG.7 and ENG.9 are combined.

*BP7: **Ensure consistency and bilateral traceability of system architectural design to the system integration test specification**. Ensure consistency of system architectural design to the system integration test specification including test cases. Consistency is supported by establishing and maintaining bilateral traceability between the system architectural design and system integration test specification system including test cases.*

Consistency of system integration testing to system architecture is, as already shown in ENG.7 BP7, primarily guaranteed through »horizontal« traceability (see also section 2.24) between system architecture elements and elements of the system integration test specification. This means that specification is needed

---

59. The original sentence is grammatically wrong. Besides, confirmation of deliverability does not take place until ENG.10.

regarding which system architecture element, including the verification criteria, is to be tested with which test case elements of the system integration tests, and vice versa. Here, too, traceability must be maintained throughout the life of the project—especially after changes are made.

BP8: Develop regression testing strategy and perform regression testing. Develop the strategy for re-testing the system elements if changed hardware items or integrated software are integrated. Perform regression testing as defined in the regression test strategy and document the results.

After changes to system elements (e.g., after defect removal or function enhancements), regression testing ensures that attributes that have already been verified by previous tests still exist (for further details see ENG.7 BP8).[60]

2.11.4 Designated Work Products

08-50 Test Specification;
See ENG.8

08-52 Test Plan
See ENG.8

11-06 System
Here the term system is understood to define the fully configured and integrated set of product elements (see figure 2–2 and related comments), e.g., the required hardware, mechanics, software, and data, including documentation.

13-22 Traceability Record
See ENG.2

13-50 Test Result
See ENG.8

2.11.5 Characteristics of Level 2

On Performance Management

System integration is an interdisciplinary activity and is usually performed by different, interacting teams/subprojects (e.g., software development, electronics development, printed circuit board layout, mechanical construction) within the overall project. Subprojects typically follow their own subproject plans which are coordinated and agreed at an overall-project level. In our experience, this is prob-

---

60. In other words: Does what verifiably worked before the change still work now?

lematic and requires particular effort, especially if subprojects are spread across different development sites. Project tracking is accordingly sophisticated. In practice, project management teams in which representatives or subproject managers from all disciplines cooperate and meet on a regular basis have proved successful (concerning the management of work products, see also the corresponding deliberations in ENG.7).

2.12 ENG.10 System Testing

2.12.1 Purpose

The purpose of the System testing process is to ensure that the implementation of each system requirement is tested for compliance and that the system is ready for delivery.

The overall system is tested, consisting of hardware, software and mechanics. In the same way as the system integration test process ENG.9 can be seen in analogy to the software integration test process ENG.7, the system test process ENG.10, with the exception of a few small details, can be considered methodologically identical to the software test process ENG.8. For this reason, the following deliberations will focus primarily on the differences to ENG.8. System testing in late phases (shortly before SOP—start of production) is sometimes called »delivery testing«.

2.12.2 Characteristics Particular to the Automotive Industry

Implementations of this process are often of distinct validation character and exceed Automotive SPICE requirements in other respects, too, since it is not sufficient only to test the system requirements. The reason for this is that in practice, system requirements specifications, as well as customer requirements specifications on system level, do not show the desired quality and completeness. Besides, in some domains, system requirements evade exact description. In vehicle development this is, for instance, true with respect to the fine tuning of chassis electronics. Instead of classical system testing, protracted test expeditions take place with test drivers working in cooperation with development engineers, tuning the chassis parameters until everybody is satisfied (see also our deliberations concerning »Characteristics Particular to the Automotive Industry« in the ENG.7, ENG.8. and ENG.9 processes).

2.12.3 Base Practices

BP1: Develop system test strategy. Develop the strategy for system testing consistent with the release strategy.

See on this point our deliberations in ENG.8 BP1. Based on the verification criteria of the system requirements, it can be determined when a test may be considered successfully passed. For functional testing, for instance, verification criteria describe the expected system behavior in detail. Ideally, development of system test cases takes place very close to requirements analysis (see ENG.2). The goal should be to achieve 100% coverage of existing system requirements with associated test cases to prove that all system requirements have been fulfilled. The release strategy with its sample phases (see SPL.2 for an explanation of sample phases in the automotive industry) is the basis of the system test strategy.

BP2: Develop test specification for system test. Develop the test specification for system test, including the test cases, to be executed on the integrated system. The test cases should demonstrate compliance to the system requirements.

NOTE: *The System testing process should start early in the system development life cycle. There is a close link to System Requirements Analysis ENG.2, System architecture design ENG.3, and Requirements Elicitation ENG.1 in developing test cases and testable requirements.*

Our deliberations shown in ENG.8 BP2 apply here, too. The requirements of the system test process should, like the requirements of the system integration test process ENG.9, already be considered during system requirements analysis for a suitable interpretation of the verification criteria. Here, too, the test strategy is documented as part of the test plan—an overall strategy across all tests is also advisable here.

BP3: Verify integrated system. Verify the integrated system against the test cases for system testing and according to the system test strategy.

NOTE: *Verification of the integrated system produces the test logs.*

NOTE: *Tests should be automated as far as possible having regard to efficiency.*

Tests are executed and their results documented in test logs. Regarding documentation of test results in test logs, see also our deliberations provided in ENG.7 BP5/BP6, and the excursus »Test Documentation« in ENG.6.

BP4: Record the results of system testing. Document the results of system testing and communicate to all relevant parties.

NOTE: *The test incident reports and the test summary report are based on the test logs.*

Analogous to ENG.7 BP6, the test log forms the basis for creating the test incident reports and the test summary report for the integrated system. Based on the test summary reports, and particularly on the results of the delivery and customer acceptance tests, a decision can be made whether system quality is sufficient to merit delivery to the customer.

> **Note to Assessors**
>
> On request of the customer, for instance, in case of urgently required defect correction for early and intermediary sample versions, system testing is partly or wholly omitted. This, and accompanying risks, must be explicitly stated in the documentation of the sample delivery. A minimum test scenario could take the regression test strategy as a guideline.

If required as a result of a defect, specifications must be revised up to system level in analogy to ENG.8 BP4.

BP5: Ensure consistency and bilateral traceability of system requirements to the systems test specification. *Ensure consistency of system requirements to the systems test specification including test cases. Consistency is supported by establishing and maintaining bilateral traceability between the system requirements and systems test specification including test cases.*

NOTE: Consistency can be demonstrated by review records.

This base practice is analogous to ENG.8 BP5, i.e., consistency between the system requirements and the system test specification is guaranteed by »horizontal« traceability (see also section 2.24) between system requirements and the system test specification. This means that for each system requirement the system test that will cover it must be defined. On the other hand, system test, too, must define which requirements it is going to test. Actual test coverage can be verified by means of reviews and corresponding review records. Here, too, traceability must be maintained throughout the life of the project—especially after changes.

BP6: Develop system regression test strategy and perform testing. *Develop the strategy for re-testing the integrated system should a system element be changed. If changes are made to system elements carry out regression testing as defined in the system regression test strategy, and record the results.*

This practice requires regression tests for the entire system (on regression tests, see ENG.7 BP8).

2.12.4 Designated Work Products

08-50 Test specification
See ENG.8

08-52 Test plan
See ENG.8

13-50 Test result
See ENG.8

2.12.5 Characteristics of Level 2

See the analogous deliberations in ENG.8.

2.13 SUP.1 Quality Assurance

2.13.1 Purpose

The purpose of the Quality assurance process is to provide independent assurance that work products and processes comply with predefined provisions and plans.

Quality assurance processes and practices are defined in several parts of Automotive SPICE, forming an (intentionally) redundant system:[61][62]

Generic Practices GP 2.2.1 and 2.2.4

In order to reach level 2 for a process, a minimum of quality assurance must be guaranteed for the process' work products, even if none of the quality assurance processes listed below are implemented.

SUP.1 Quality Assurance

SUP.1 is the basic quality assurances process at the project level. In contrast to the generic practices, not only the quality of the work products is addressed here but also the quality of processes. SUP.1 coordinates and monitors the activities of SUP.2, SUP.3 and SUP.5.

---

61. Even if an organization has not taken any processes of the SUP-processes mentioned into consideration, QA practices are nevertheless required at Level 1 by the engineering processes. At Level 2 they are required by the generic practices GP 2.2.1 and 2.2.4.
62. Some of the processes mentioned here are not included in Automotive SPICE but can, if required, be adopted from ISO/IEC 15504.

SUP.2 Verification

SUP.2 systematically ensures that each work product of a process or of a project adequately reflects the prescribed requirements. This includes that verification criteria are identified, that defects are detected, recorded, and tracked, and that the results are made available to the customer and other involved parties.

Verification Activities in ENG.6–ENG.10

In these engineering processes, numerous verification activities that are closely involved in the development activities are required for the work products. Quality assurance monitors the performance of these activities even if it does not always carry them out itself. These activities, which ought to be performed very soon after the work products have been created, may prevent subsequent faults or reduce verification costs in later phases. These verification activities are already applicable at level 1 and if none of the other quality assurance activities mentioned here have been implemented.

SUP.4 Joint Review

SUP.4 focuses on the synchronization of those involved in the project and on meeting their interests and needs. The joint reviews with involved persons and groups (»stakeholders«) specified there include management reviews and technical reviews. Particularly the latter represent quality assurance measures.

SUP.5 Audit

SUP.5, similarly to SUP.1, ensures that work products and processes satisfy quality requirements. The crucial difference is that audits must be performed by an independent party.

MAN.4 Quality Management

MAN.4 requires a quality management system which is to systematically ensure customer satisfaction through monitoring of product and service quality. MAN.4 includes both project and organizational level and forms a kind of umbrella for all the other quality assurance activities.

2.13.2 Characteristics Particular to the Automotive Industry

In the development of hardware and software systems, required technical domain know-how is often so specialized and requires such extensive experience, that the definition and implementation of quality requirements on technical work products (e.g., brake and chassis systems control, ECUs) can only be ensured by domain experts. Due to its interdisciplinary nature, and particularly in large organizations, quality assurance is typically performed by different expert groups.

2.13 SUP.1 Quality Assurance

The practical implementation of the SUP.1 process must therefore ensure the coordination of all involved individuals.

Classic quality issues, defined and mapped, for instance, in ISO 9001, EFQM or ISO/TS 16949, have been known in production for many years and are well established in organizations. New methods are employed to improve production even further, although only small improvements can be expected here in the future. Unfortunately, quality assurance issues are not always granted the necessary and appropriate weight or status in development, and required staffing or infrastructure is not always sufficiently provided. In most cases, small organizations with up to 50 developers have only one engineer to ensure developmental quality assurance, and besides his QA tasks he often remains involved in development or testing activities.

Trying to apply the »classic« QA approaches in development will usually not yield the expected results, because methods, proceedings and content differ from each other. To clearly and comprehensibly present these differences in the organization, and to obtain management commitment for the support of quality assurance in development, is an important precondition for appropriate implementation of this process.

2.13.3 Base Practices

BP1: Develop project quality assurance strategy. *A project level strategy for conducting quality assurance is developed. This strategy is consistent with the organizational quality management strategy.*

NOTE: *The quality assurance process may be co-ordinated with the related SUP.2 Verification, SUP.4 Joint review, Validation and Audit processes.*

Such a strategy is typically documented as part of a quality plan (see BP3). The strategy determines the basis which is then detailed within the plan. If other QA processes are implemented (e.g., SUP.2, SUP.3, SUP.4, SUP.5), they must be included and coordinated within the context of this strategy.

It makes sense (but is not required below Level 3) to define such a strategy for the whole of the organization. The organization can then define minimum requirements for quality assurance measures in all its projects and describe their inclusion in the overall system of higher-level activities. The following list shows some of these requirements:

- Statement of the organization's quality goals in the company's quality policy
- Compliance with internal and external quality requirements (which may be derived from standards)
- Description of the general procedure regarding the performance of QA activities in projects (including responsibilities)

- Description of tasks needed to coordinate all relevant QA /QM activities, and of their interaction (integrated management system)
- Acceptance of the customer's quality standards as a contractual element of (software) development
- Communication to all affected staff about the quality strategy
- Existence of a quality plan as a basis for other plans (e.g., verification)
- Detailed description of QA measures in accordance with the QA strategy

The organization's quality policy should then be reflected in the project's development activities. The coordination of all QA/QM activities can be supported by means of an integrated management system comprising work safety, environmental protection, quality assurance for series production and for development.

BP2: Develop and maintain an organisation structure which ensures that quality assurance is carried out and reported independentely. Quality assurance team members are not directly responsible to the project organisation—they work independently from it.

The organizational structure must support the work of quality assurance in projects. To do so it is necessary for quality assurance staff to be able to work independently from project-internal constraints. This is typically ensured by an organizational unit which is independent from development and to which QA staff members report technically and administratively. Criteria for independence are as follows:

- The existence of an independent organizational structure that is internally known and visible to the external world
- A defined, independent reporting system up to the highest executive level of the organization

> **Note to Assessors**
> In some projects, the role of the quality assurance manager is assigned to a developer in addition to his primary tasks, e. g., coding or testing. Organizationally, he still reports to the project manager. This is not in compliance with Automotive SPICE, since in such cases a conflict of interest usually exists between development and QA tasks.

2.13 SUP.1 Quality Assurance

BP3: Develop and implement a plan for project quality assurance based on a quality assurance strategy

NOTE: Quality assurance plan[63] may contain quality assurance activities, a schedule of activities, assigned responsibilities, resources required, guidelines and quality standards for requirement, design, coding and testing work products.

The quality plan is the central document for the planning of a project's quality assurance activities. Without written documentation it is hardly plausible that required activities will actually and promptly be performed or that quality assurance requirements be complied with. If other SUP.1 related processes are implemented, these must be included in the plan and all activities must be jointly coordinated. Updates and adjustments to the quality assurance plan are done either regularly or on demand. As far as content is concerned, the minimum requirements for a quality assurance plan are as follows:

- What is going to be evaluated?
  Which work products and processes are to be checked? It is unrealistic in a project to review *each* work product and *each* process substep. A reasonable selection must therefore be made in accordance with the project's circumstances. In doing so one may, for instance, determine that certain central issues will be examined during each evaluation, others only randomly, and others not at all.
- Who performs the evaluations and which resources are required for the job?
  A decision is made that determines which organizational unit or role is responsible for the respective check, and what qualifications are necessary. In a project, staff members are usually assigned to individual activities by name. In addition, staffing and infrastructure resources are planned.
- What will be checked?
  What is the basis for each respective check?
    * In case of work products, this may include a content-related check (for instance, in case of a code module, against design and requirements) and formal criteria (for instance, compliance with coding guidelines), but also standards, legal requirements, etc.
    * Processes are usually checked against the process description
- How are evaluations to be carried out, and what will be done to stop deviations?
  What is the evaluation method, and how are results documented? What is done to ensure that deviations are removed?
- What is the evaluation frequency?
  To fulfill the process, regular checks (for instance, at »Quality Gates«) must be performed. There are no fixed rules regarding frequency. A popular strat-

---

63. The corresponding work product 08-13, however, is called a »quality plan«.

egy, for instance, is to make the frequency dependent on process conformity (i.e., if process conformity in a project or a part of the organization is good, the frequency is reduced). Checking usually takes place at defined milestones in the project, for instance, prior to customer deliveries. In an assessment, the question of frequency is evaluated in the light of the project's and organization's circumstances to see if the process purpose has been accomplished.

- Which consequences result from the evaluation?
  Will it be possible to start the next phase of the project, perhaps only with conditions imposed, or are specific measures necessary if quality assurance declares the evaluation failed? In case of identified non-fulfillment of requirements, an appropriate procedure must be specified.

The staffing and infrastructure effort required for the implementation of the activities of the quality plan must be planned. This is either done in the quality plan itself, in the overall project plan, or in the subproject plans. If planning is distributed over several documents, the plans must be kept consistent with each other.

Fulfillment of requirements derived from the quality assurance plan is ultimately a criterion by means of which quality assurance grants its approval, for instance for deliveries to the customer or not. Publication of the plan in the organization is necessary to ensure that everybody involved in the project knows its content and the chosen course of action.

BP4: Maintain evidence of quality assurance. *Define and maintain the records that demonstrate that planned quality assurance activities have been implemented.*

It is defined which QA records are created in the project; these are kept up-to-date. Among others, the following are included:

- Definition of which record is created for which QA measure/activity
- Proof that QA activities were performed according to plan

The assessment checks the evidence as to whether or not QA activities have been performed. These documents are an indicator for their implementation and are sample-checked. At this point, in particular, activities are only considered to be performed if they have been documented. Projects should define which records are only logged once and which are to be updated. Different types of records are possible.

2.13 SUP.1 Quality Assurance

BP5: Assure quality of work products. Carry out the activities according to the quality assurance plan to ensure that the work products meet the quality requirements.

NOTE: *Product quality assurance activities may include reviews, audits, problem analysis, reports and lessons learned that improve the work products for further use.*

NOTE: *Non conformances detected in work products may be entered into the problem resolution management process (SUP.9) to document, analyze, resolve, track to closure and prevent the problems.*

According to BP1 (strategy) and BP3 (quality plan), work products are evaluated and results documented. Quality assurance in the project is usually not responsible for carrying out technical checks (e.g., tests) itself but to ensure that the activities defined in the quality assurance plan are done according to schedule, that the expected results are achieved, and that compliance with requirements can be evidenced.

In order to prove compliance, work products are usually examined together with the project team. Despite our deliberations in the previous paragraph, it helps if quality assurance staff members have domain know-how (e.g., of the engineering processes), even if tests or project-internal reviews[64] are performed either by the developer himself or by experienced, project-independent test department staff. The final evaluation whether work products were developed in compliance with the requirements or not should be carried out by quality assurance within the context of a release (e.g., at delivery or at important milestones). Formal[65] proof regarding compliance (between result and requirement) is provided by QA.

In order to provide proof that quality assurance measures satisfy the requirements as planned, it is hardly possible to inspect 100 percent of the results. Instead, an intelligent and flexible spot-check strategy with high coverage is needed. »Adjustments« to this strategy are stated in BP3 (What is reviewed by whom? How intensive is the check? How often are reviews carried out?) and are based on the degree of confidence one has in the project. A review prior to delivery to the customer should be more comprehensive and more detailed than that of an internal work product.

---

64. On established verification methods see our deliberations in SUP.2.
65. Proof regarding content may be provided by others because, among other things, special domain know-how is usually needed in such cases.

> **Note to Assessors**
>
> Identified deviations relating to work products must be tracked and dealt with until they are corrected (for instance, by means of the problem resolution management process (SUP.9)). In order to prevent re-occurence, the organization should provide appropriate, preventive means and methods.

BP6: Assure quality of process activities. *Carry out the activities according to the quality assurance plan to ensure assurance that the processes meet the defined requirements of the project*

NOTE: Problems detected in the process definition or implementation should be entered into the process improvement process (PIM.3) to describe, record, analyze, resolve, track to closure and prevent the problems.

NOTE: Process quality assurance activities may include process assessments and audits, problem analysis, regular check of methods, tools, documents and the adherence to defined processes, reports and lessons learned that improve processes for future projects.

NOTE: In case of supplier involvement, the quality assurance of the supplier should cooperate with the quality assurance of the customer and all other involved parties.

Process audits are performed and documented according to the methods defined in BP1 (strategy) and BP3 (quality plan). To ensure that the processes applied in the project adequately satisfy the defined requirements, process compliance audits are performed. SUP.1 is the only place in Automotive SPICE that requires such checks. SUP.1 is therefore particularly important because such audits ensure that the work sequences and procedures[66] required by the organization are actually implemented in projects. Process compliance audits cannot be performed by staff that are directly involved in the project, other project team members (not directly involved in this process) are also not acceptable in this role. Usually, project-independent staff (e.g., QA staff) are assigned for this. It is common practice during audits to interview project team members and also to review work products.

The definition of process requirements for a project is usually determined by management instructions stating which processes are binding in which organizational units and project types. Deviations often need the approval of QA. At Level 3, there are additional deviations in the form of tailoring. In this case, tailoring guidelines specify the room to maneuver at project level. Formal proof regarding conformity of process requirements and process implementation in the project is provided by QA.

Identified deviations from processes must be dealt with and tracked to closure. In order to prevent re-occurence, the organization should provide appropri-

---

66. Usually expressed by project descriptions.

ate, preventive means and methods, e.g., by means of an (S)EPG ((Software) Engineering Process Group). During the assessment, examples can be collected to learn if and how identified deviations lead to changes in the process definition or implementation. In doing so, attention should be paid to the consistency and traceability of examined data and information.

Assessing process implementation in the project, external process interfaces (e.g., to other subprojects or to the customer) must also be considered and evaluated accordingly. In our experience, many problems in a project arise from the fact that interfaces in process definitions are insufficiently described and accounted for and that they are therefore only rudimentarily implemented.

In one of its notes, Automotive SPICE recommends direct cooperation between all involved QA organizations. It is advisable to jointly determine, plan, and perhaps also perform quality assurance measures to reduce interface losses. It also helps to ensure that all stakeholders are kept at the same level of information and to avoid duplication of work. At the same time, the internal effects of quality assurance and its independent reporting path within the organization should be exploited as needed to achieve fast resolution of problems.

BP7: Track and record quality assurance activities. Records of quality assurance activities are produced and retained.

Quality assurance activities are tracked, monitored, and recorded. On the one hand, this ensures that quality assurance in projects takes place as planned; on the other, that quality assurance with all its tasks remains subject to a certain amount of control and does not become a decoupled institution. Responsibility for monitoring quality assurance lies with management.

The [ISO 9001] defined »management responsibility« in section 5.6.1 as follows: »Top management shall review the organization's quality management system, at planned intervals, to ensure its continuing suitability, adequacy, and effectiveness« The same applies to quality assurance in development projects.

Tracking of QA activities, i.e., whether they are carried out (in a timely manner), is typically done by a management function within the QA organization itself. Additional external audits may be held to ensure the effectiveness and appropriateness of the tracking mechanism and of the QA system as a whole.

The performance of all quality assurance activities must be verifiable. As far as documentation of QA activities is concerned, document management requirements apply, e.g., regarding versioning (see also DIN 9001 and SUP.7). The following list provides some examples:

- Proof that QA activities have been fully performed and according to plan (e.g., prior to important milestones)
- Complete records must be available
- Records must be reproducible
- Specifications must be available as to how records are to be archived and reproduced

BP8: Report quality assurance activities and results. *Regularly report performances, deviations, and trends of quality assurance activities to relevant parties for information and action.*

NOTE: The quality assurance may use an independent path to report regularly the results to the management and other relevant stakeholders.

Quality assurance regularly informs all relevant parties/persons on activities and results of QA measures, including the following:

- Reporting interval and distribution list are specified.
- Information is published.
- Reporting of deviations from prescribed quality goals (e.g., providing metrics regarding defect development in the project over time to allow, for instance, estimations on how long it will take to correct all defects.)
- Reporting on performed QA or corrective measures in the projects
- Regular reporting to management

One of the premises for the reporting of activities is that quality assurance activities are actually recorded (see BP7). Reports are sent to staff that are directly involved in the project and to the executive level, (S)EPG and, if required, further stakeholders.

Figure 2–10 provides an example of a quality report. It includes a quantitative overview of the completion status of activities resulting from QA findings (left scale, vertical bar) and a process conformity rate (right scale in percent, horizontal lines) as a result of process compliance checks. The latter illustrates the percentage of all practices of the processes implemented according to the specifi-

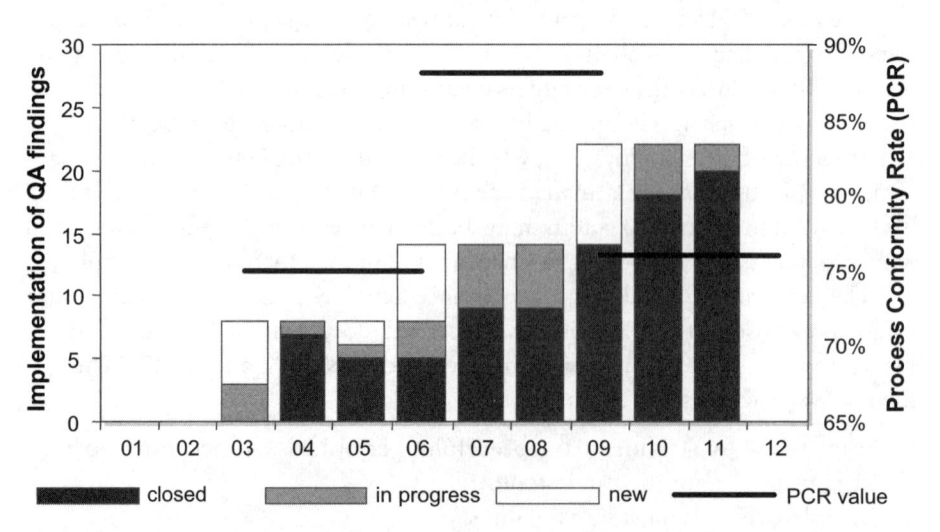

Figure 2–10 Example of a QA report to management: Completion status of activities resulting from QA findings and process conformity rate

cations and the number of practices that have not yet been implemented. Tracking is visible over 12 months of a year: the vertical bars in the figure show the completion status of activities resulting from QA findings (e.g., detected problems/defects) accumulated in absolute figures. The horizontal lines in the figure show the results of three process compliance checks in the form of a process conformity rate.

BP9: *Ensure resolution on non-conformances.* *Deviations or non-conformance found in process and product quality assurance actitvities should be analyzed, corrected and further prevented.*

Identified deviations are reliably and completely resolved. For this purpose the auditor discusses the deviations with the relevant members of staff and agrees on measures for their resolution. Monitoring the resolution is the responsibility of quality assurance and is usually done by the auditor. If resolution is not achieved as agreed, escalation is initiated (see BP10).

Furthermore, the root causes of non-conformities and deviations are analyzed to prevent re-occurrence. Quality assurance must collect and evaluate information for the purpose of analysis, correction, and avoidance. Typical evaluations are root cause analyses (»What was the reason?«) and statistical analyses (»What are frequent causes of deviations?«). Measures can be derived from the results to remove the causes either in the project or the organization as a whole. Typical measures at an organizational level are, for instance, the removal of know-how deficits through training or improvements of methods, processes, tools, templates, checklists, etc. Measures at an organizational level are expediently treated by the PIM.3 process (process improvement); measures at a project level by the SUP.9 process (problem resolution management).

In practice, project related measures concerning processes can in most cases only be accomplished with organizational project support and/or by measures that apply to the entire organization. Examples from actual development projects show that in many cases, this issue is not paid appropriate attention. Defects are repeated, the learning curve of new projects very often starts from scratch and measures that proved helpful in predecessor projects are forgotten with the new project team and different project conditions. It is especially for this reason that the automotive industry requires preventive measures aimed at defect avoidance regardless of the maturity level.

BP10: *Implement an escalation mechanism.* *Develop and maintain the escalation mechanism that ensures that quality assurance may escalate problems to appropriate levels of management to resolve them.*

If non-conformities/deviations are not removed or cannot be resolved in the project itself, management must be informed to support or bring about a solution. Among others, it includes the following:

- Definition/description of escalation mechanisms/levels (right up to the executive level)
- Proof that escalation mechanisms work
- Definition of the responsibilities and authorities of the individual escalation levels (addressor/addressee)

> **Note to Assessors**
>
> An organizational chart as proof of the escalation path is usually not sufficient, except in cases where the respective organization is very small. In the assessment, appropriate samples of evidence, for instance, correspondence items, need therefore be presented to prove that the escalation mechanism works.

2.13.4 Designated Work Products

08-13 Quality Plan[67]

The quality plan contains the methods and measures listed in BP3. From Level 2 onward, it also includes some of the planning elements required in PA 2.1.

> **Exemplary Structure of a Quality Plan**
>
> 1. Area of Application
> 2. Responsibilities and Authorities of QA Staff
> 3. Required Resources
> 4. Budgeting of QA Activities
> 5. Quality Assurance Measures and Schedule
> 6. Standards and Procedures to be Applied during Auditing
> 7. Procedures for Deviation Handling (Including Escalation to Management)
> 8. Documentation to be Created
> 9. Feedback Procedure to the Project Team

13-07 Problem Record

On this point, refer to SUP.9. Problem records can be used to document the deviations identified in BP3. Problem records are sometimes combined with quality records (see WP-ID 13-18). Figure 2–12 provides an example of a problem record that has been integrated into a quality record.

---

67. In the Base Practices, the quality plan is referred to as »quality assurance plan«. This is an inconsistency in the model.

13-18 Quality Record

The quality record documents the verification measures required in BP5 and BP6. Figures 2–11 and 2–12 illustrate an example of such a report. It sometimes also contains the contents of problem records and corrective actions in list form (see figure 2–12).

QA Conformity Review	
Project:	
Review Date:	
Subject:	
Reviewer:	
Participants:	
Results:	
○ Deviations (Yes/No):	
○ Number of deviations:	
Status of implementation of measures:	
Summary of Results:	

Figure 2–11   *Example of a Quality Record (Cover Page)*

Deviations and measures										
ID	Evaluated object	Overall impression	Number of deviations	Deviation in detail	Measure ID	Measure in detail	Date measure was initiated	Responsibility for measure	Target date	Status, comments

Figure 2–12   *Scheme of a Quality Record (Measures Page)*

14-02 Corrective Action Plan

Listing all corrective actions in a table serves to find a solution for the issues addressed in the problem records, i.e., responsible staff, solution proposal (list of measures for problem resolution), deadlines (start and completion date of the corrective measure), status and subsequent checks must all be specified. In practice these details are typically documented in the quality record (see above).

> **Note to Assessors**
>
> Above all, the following needs to be checked in this process:
> - Were sufficient reviews planned?
> - Did planned reviews actually take place?
> - Are verification results documented?
> - Were measures defined sufficiently and in time?
> - Were measures successfully performed and accepted?
> - Can adequate reporting activities be evidenced?
> - Did escalations result in the solution of problems?

2.13.5 Characteristics of Level 2

On Performance Management

Planning of quality assurance activities is primarily done in the quality plan, whereby individual checking activities, deadlines, and executing staff are typically reflected in the project plan. Activity tracking is usually done on three levels:

- In principle, the project manager tracks and controls all QA activities except the process conformity checks[68].
- Process conformity checks are usually conducted by project-independent QA staff. The responsibility to track that these checks have been performed in sufficient number and in the proper way typically lies with a manager with quality assurance.
- Management is responsible for monitoring the quality status of the overall project, for instance, by means of indices and charts included in the regular status reports.

Quality assurance at project, organizational, and varying maturity levels

Process compliance checks are an essential element of effective quality assurance. However, these checks require project-independent (QA) staff and therefore an appropriate organizational structure. Proper quality assurance is not feasible with a project's own resources alone. A second precondition for the implementation of process compliance checks is that the processes and requirements against which checking is carried out must be defined. At level 1 this is usually only established in a very rudimentary way. Because of the nature of such process compliance checks, a standardized QA process is normally needed that requires elements of Level 3.

The bottom line is that SUP.1 requires elements of Level 2 and 3 and that certain organizational prerequisites must be met. In assessments, this may lead to

---

68. Since project management itself is subject to process compliance checks.

the problem that—without such elements—the accomplishment of the process purpose at Level 1 can be evidenced only with difficulty or not at all.

Quality assurance regarding quality assurance

At Level 2, some sort of »quality assurance of quality assurance« is implemented by GP 2.2.4. This means specifically that the implementation and effectiveness of the process is periodically audited. To do so, central SUP.1 work products, for instance, the quality plan or samples of deviation reports, are examined by an independant body. This may consist of external auditors or of persons out of the QA organization with sufficient distance to the operative level of SUP.1.

2.14 SUP.2 Verification

2.14.1 Purpose

The purpose of the Verification process is to confirm that each work product of a process or project properly reflects the specified requirements.

Among experts, verification is usually understood to be a process that is specific to individual development phases (see, for instance, [Spillner et al. 2005]) during which the correctness and completeness of work products of a particular phase are checked against the direct requirements[69] on the respective work products. During verification of a software component, for instance, the emphasis is placed on checking against the component's design specifications and against coding guidelines. In a way, verification answers the question »Am I building the system right?«, i.e., does it comply with the requirements?

Validation, on the other hand, answers the question »Am I building the right system?«, i.e., is it of practical use? Validation therefore focuses on whether a system is suitable for its intended purpose. This is done primarily by testing the system in practice and secondarily by testing against the customer and system requirements.

Most of the established methods may in practice be used for both, verification and validation purposes. However, methods differ:

- For verification, reviews, walkthroughs, tests at different levels, and static code analyses are primarily used.
- For validation, the predominant method is practical trial, but also black-box testing or a variety of special tests like stress tests, performance tests, and load tests. Simulations are also suitable for early validation.

---

69. In the process purpose this is called »requirements«. However, this means requirements in the general sense of the word, not only the original customer requirements.

Automotive SPICE describes verification and validation activities on numerous occasions (see also the overview provided in SUP.1), among others in the engineering processes as well as in Process Attribute PA 2.2. If the SUP.2 verification process is implemented it may exceed the former in the following points:

1. SUP.2 describes additional measures running parallel to development, and performs joint planning and coordination for these measures.
2. SUP.2 activities enjoy a certain degree of independence from development activities.

As already mentioned, SUP.2 is closely linked to the validation process[70] (ISO/IEC 15504-5 SUP.3). Methods to be applied and process methodology are very similar. That is why in many process implementations verification and validation activities are closely coordinated, for example in the form of a joint verification and validation plan (»V&V plan«).

Verification is an incremental process running right through the entire project and starting with requirements verification. This is then followed by the verification of work products and different product prototypes created in subsequent development phases.

2.14.2 Characteristics Particular to the Automotive Industry

In the automotive industry, verification activities require very specific and detailed know-how and experience. It therefore suggests itself that these activities are carried out either by the developers themselves or by specially trained and experienced domain experts.

2.14.3 Base Practices

BP1: Develop a verification strategy. *Develop and implement a verification strategy, including verification activities with associated methods, techniques, and tools, work product or processes under verification, degrees of independence for verification and schedule for performing these acticvities.*

NOTE: *Verification strategy is implemented through a plan.*

NOTE: *Software and system verification may provide objective evidence that the outputs of a particular phase of the software development life cycle (e.g., requirements, design, implementation, testing) meet all of the specified requirements for that phase.*

---

70. Because of the assessment experience of the members of the Automotive SIG, the validation process SUP.3 was not implemented as an automotive specific process in Automotive SPICE. Should the validation process be required in an assessment, the process as described in ISO/IEC 15504-5 is to be applied.

2.14 SUP.2 Verification

NOTE: *Verification methods and techniques may include inspections, peer reviews (see also SUP.4), audits, walkthroughs and analysis.*

As a matter of principle, the verification methods to be applied are left open in SUP.2. Typical methods applied in practice are, for instance:

- Test
- Static code analysis
- Reviews, inspections, walkthroughs
- Simulation
- Traceability analysis
- Assessments[71], audits

The methods mentioned here are partly suited for the verification of work products, partly for the verification of central, quality-relevant processes (e.g., in the form of configuration audits for the configuration management process or its work products), and partly for both.

Each of the groups of methods mentioned here has its separate profile regarding the area and time of application, and the type of defects that are typically detected with it. The verification strategy should therefore specify a reasonable, holistic approach with activities, associated methods, intended times of application within the lifecycle model, and a concrete schedule. It may form the overall framework for the test strategy required in the ENG processes.

In practice, it is not possible and does not make sense to verify *each* work product and a reasonable selection must therefore be made[72]. A risk evaluation can also be used for the selection, for instance, to verify critical system parts. Another important criterion is that early lifecycle work products should be a first choice if this helps to avoid later defects (e.g., by reviewing system requirements against customer requirements). Other important elements of the strategy are:

- Specification of the verification methods to be used, for which work products, in particular phases of the project
- Specification of the verification environment (test drivers, tools)
- Prioritization of the verification activities, if because of time constraints it is not possible to perform all of them
- In incremental development, definition of the different verification intensity during individual increments (e.g., test depth)

---

71. Assessments in the general use of the term, not necessarily limited to assessments in terms of Automotive SPICE.
72. Unfortunately, this is in contrast to the unrealistitic wording given in the process purpose (»each work product«).

> **Note to Assessors**
> An assessment should ascertain that the existing strategy is continuously adapted to changing project constraints. This requirement is specified in the process outcomes, and therefore mandatory, but was forgotten in BP1.

If the SUP.2-process and the engineering processes (which also include verification activities) are implemented in the organization, the strategy must also include an agreement on the different evaluation activities. In verification, a certain degree of independence[73] of verification activities from development activities is usually aimed at. Three parameters can be distinguished (see also [IEEE 1012]):

- Technical independence: Verification activities are performed by other staff than those directly involved in development (also known as the »four-eyes-principle«). The advantage is that this avoids »professional blindness«.
- Organizational independence: Verification activities are located in a separate organizational unit under their own management. The advantage is that verification can be carried out without restrictions or pressure coming from development.
- Financial independence: Verification activities have their own budget independent from development. The advantage is that development cannot restrict verification through cuts or rededication of budget.

Regarding these three parameters, different combinations are possible and common, ranging from the classical form (independence regarding all three parameters) right up to an embedded form, i.e., development and verification are under joint management, have a common budget and a joint pool of staff.

BP2: Develop criteria for verification. *Develop the criteria for verification of all required technical work products.*

Verification criteria specify which conditions must be met in order for a technical work product (for instance, a vehicle component) to be considered successfully verified. They depend on the type of the work product and the verification method to be applied. The criteria result from the requirements which are to be met for the respective work product. These requirements are mainly constituted by the outputs of the immediately preceding development phases. Customer requirements, for example, are input for product requirements, product requirements are input for product design, product design is input for coding, etc. Additional verification criteria usually arise from development processes to be com-

---

73. Automotive SPICE does not provide any concrete requirements, only the degree of independence must be specified.

2.14 SUP.2 Verification

plied with and from applicable standards and legal requirements. Verification criteria usually require further detailing, for example:

- Each customer requirement must be reflected in at least one product requirement.
- To what degree will it be necessary to use development independent staff for reviews.
- Criteria regarding test cases, e.g.,:
  - All inputs are to be tested using correct and incorrect values.
  - All inputs are to be tested using extreme and normal values.
  - Test cases are to be designed in such a way that they generate normal and extreme values for all outputs.
  - For each system function there must be at least one test case.
  - For each menu item there must be at least one test case.
- Test coverage measurements: During white-box testing, specific rules are to be observed regarding test case completeness (see excursus »Derivation of Test Cases« in ENG.8).
- Checklists are to be used during reviews.
- Static analysis tools are to be used for code.

> **Note to Assessors**
>
> The criteria for the verification of work products differ from the verification criteria found in the ENG processes (on this point, see section 2.24): a verification criterion always refers to one individual requirement only. In contrast, criteria for the verification of work products refer to the success criteria for the complete verification step.

BP3: Conduct verification. Verify identified work products according to the specified strategy and to the developed criteria to confirm that the work products meet their specified requirements. The results of verification activities are recorded.

Each verification is performed according to the strategy and the defined verification criteria, and the results are recorded (see excursus »Test Documentation« in ENG.6). This could, for instance, mean that written and summarized evidence exists stating that all software units modified during a development cycle have been 100% reviewed using project-specific checklists, that results are documented, and that all findings have been adequately resolved.

BP4: Determine and track actions for verification results. *Problems identified by the verification should be entered into the problem resolution management process (SUP.9) to describe, record, analyze, resolve, track to closure and prevent the problems.*

During verification, identified problems are recorded in a problem record and passed on for systematic problem removal. This is ideally done in the form of the problem resolution management process SUP.9, if it is implemented. Strictly speaking, the application of certain verification methods (e.g., tests) often does not detect defects as such but rather deviations between observed behavior and expected behavior. Often causal analysis is needed here to see if the observed deviance can actually be traced back to a defect or if there may be other reasons (e.g., inaccurate specification, false values resulting from test methodology, external influences in the test environment caused, for instance, by sensors or other systems connected to product under test). Root cause analysis can turn out to be very labor intensive, for instance, if the deviating behavior occurs sporadically and cannot be reliably reproduced.

BP5: Report verification results. *Verification results should be reported to all affected parties.*

In the simplest case, involved parties are directly informed, for instance, if in a review the author of the work products is present. In any case, each involved party must receive a report or minutes stating the verification results. The report can have different forms (see excursus »Test Documentation« in ENG.6):

- For the developer: Test incident report, test log
- For the project management: Test summary report
- For the customer: Delivery documentation

2.14.4 Designated Work Products

13-07 Problem Record

see SUP.1

13-25 Verification Results

Verification results must identify or contain at least:
- Objects that have passed verification
- Objects that have failed verification
- Objects that have not yet been verified
- Problems detected during verification
- Perhaps results of a risk analysis
- Recommendations concerning measures
- Conclusions drawn from the verification

14-02 Corrective Action Register

see SUP.1

19-10 Verification Strategy

The verification strategy describes the verification methods, procedures, and tools needed for the implementation of the individual verification steps as well as the work products and processes to be verified. Moreover, it states the degree of independence granted to verification and a schedule for the implementation of verification activities. Based on known constraints and risks, additional verification criteria are described, for instance, criteria for test start, test completion, test abortion, and test resumption.

2.14.5 Characteristics of Level 2

On Performance Management

Planning and tracking of verification activities depend on whether and to what extent these are carried out by project or project-external staff. In the first case, planning and tracking are done by means of the project management methods applied in the project. Otherwise a separate plan is usually drawn up that is synchronized with the project plan and tracked by project-independent staff.

On Work Product Management

Due to its particular importance, priority attention should be given to reviewing the verification strategy.

2.15 SUP.4 Joint Review

2.15.1 Purpose

The purpose of the Joint review process is to maintain a common understanding with the stakeholders of the progress against the objectives of the agreement and what should be done to help ensure development of a product that satisfies the stakeholders. Joint reviews are at both project management and technical levels and are held throughout the life of the project.

A review is understood to be the checking of review objects (e.g., development results in the form of documents or code) against requirements and applicable guidelines by suitable checkers. The goal of this check is to find defects, weaknesses and gaps in the review object, to comment and document them, and to establish the maturity level of the object.

The term review has a very different meaning in literature compared to practical usage. Automotive SPICE does not go into any details regarding the different review methods like inspection, walkthrough, »desktop test«, peer review, and so on. [IEEE 1028] can be taken as a reference here, and further information is provided, for instance, by [Gilb 1993] and [Freedman et al. 1990].

Especially in a project's early phases, reviews are an effective and cheap way to check the quality of concepts or customer requirements, and to establish a common problem-understanding in the project. Reviews in later development phases check development results like design documents, code, or test cases. SUP.4 addresses the following reviews:

- (Project) management reviews: A systematic investigation with management participation (project management or line management), perhaps with the participation of customer representatives. Here the review object is typically the project status (project progress checking) or important development results are checked and aligned.
- Milestone reviews: At important project milestones, the project status and the development results to be completed at these milestones are checked against defined criteria, typically with management participation (see previous bullet).
- Technical reviews: Checking of development documents by experts with a focus on technical issues. Examples according to SUP.4 are hardware resource consumption, new requirements, and the application of new technologies and associated development results.

Reviews can be performed with little or great effort. A relatively elaborate form are »Inspections«, which follow strict rules. Inspections allow checkers advance evaluation of the review object and are chaired by an assigned facilitator who collects the results in the inspection meeting (see also [Gilb 1993]). Such methods are sometimes referred to as »formal reviews«. Other methods do without one (or more) of the components (e.g., no entry check of the review object by the facilitator, no prep time for checkers, no meeting but written comments instead) or cost-driving forces (no facilitator role, smaller number of checkers, e.g., reduced to author plus one colleague). In each case the individual steps and responsibilities must be accurately defined in the review process. If required, support materials (e.g., checklists and templates) must be available.

Each of the mentioned methods has its own justification and its own pros and cons. Long-term surveys have shown, for instance, that most of the problems are found by means of inspections. Inspections are therefore very well suited for vital documents such as requirements documents, architecture, design, and so on. In the case of many small reviews (e.g., for code), a meeting between author and colleague may sometimes be sufficient. In practice, most organizations possess a whole spectrum of review processes for different, application-specific purposes.

From an Automotive SPICE point of view, however, attention should be paid to the fact that rating-relevant evidence regarding reviews must be available. As a minimum, the following is recommended:

- Planning, identity of the involved individuals, procedure regarding distribution of the review objects, deadlines, etc., are traceable.
- Review results are documented.
- Problem resolution can be verified.

2.15.2 Characteristics Particular to the Automotive Industry

None

2.15.3 Base Practices

BP1: Define review elements. *Based on the needs of the project, identify the schedule, scope, and participants of management and technical reviews; agree all resources required to conduct the reviews (this includes personnel, location and facilities); establish review criteria for problem identification, resolution and agreement.*

As part of the review planning the following needs to be determined: review objects, participants and required resources, schedule (e.g., review phases, i.e., necessary preparation times, maximum duration of a review and follow-up times), locations, perhaps necessary reference material and appropriate guidelines (e.g., review process, checklists).

Prior to the review, review criteria must be determined against which the review object will be verified. Review criteria for problem identification must indicate if and when a problem exists. In technical reviews, defect classes may be specified. These may be severe defects (»majors«, e.g., content-related problems) and lesser defects (»minors«, e.g., spelling errors). Here, review criteria must indicate how these classes are defined. Moreover, a definition must be provided that helps to pinpoint exactly when there is a problem (»deviation against what?«).

Such criteria must be available in the form of rules, standards, input documents, customer requirements, process requirements (for instance, content-related and structural requirements on documents created within the context of the process' performance). It is also advisable to compile these criteria into checklists for the most important review objects. Review criteria for problem resolution must specify what needs to be done for the resolution of problems, for instance, that a serious defect found in a technical review needs to be removed by the author within *n* days and that the removal is to be verified by the review facilitator. Review criteria for the agreement of problems must define how the agreement process works among the reviewers (e.g., vote or consent).

BP2: Establish a mechanism to handle review outcomes. Establish a mechanism to ensure that review results are made available to all affected parties; establish a mechanism to ensure that problems detected during the reviews are identified and recorded; establish mechanisms to ensure that action items raised are recorded for action.

This and other requirements from other SUP.4 Base Practices are normally not specified individually for each project but expediently regulated for the entire organization in the form of guidelines or process descriptions[74]. A process description may, for instance, specify the following contents:

- Roles in the review process (usually review facilitator, checker, minute-taker, author)
- Review planning content (review object, deadlines and location, checkers, checking criteria, entry and output criteria for the review)
- Implementation of the reviews (individual preparation of the checkers, review session, procedure in case of distributed reviews: for instance, use of video conferences)
- Documentation of the review results (passed/ not passed, defect list with action points, next steps)
- Problem classes (e.g., »majors« for critical defects, »minors« for non-critical/formal defects) as well as review criteria for the purpose of result classification: What is the time needed for the resolution of individual classes (e.g., critical findings are resolved immediately, non-critical in the next version); what is the critical number of findings at or above which a review is considered failed and must be repeated, e.g., » three critical findings«?
- Distribution and communication of review results
- Check of revision completion
- Check of review effectiveness and efficiency

In addition to a defined review process, the following would be available in a Level 3 organization:

- Accompanying training for review leader and checker
- Support templates for invitation, review minutes, defect list, etc.
- Checklists with review criteria
- Support by review experts

Figure 2–13 provides an exemplary overview of the review process.

---

74. Not required below Level 3.

2.15 SUP.4 Joint Review

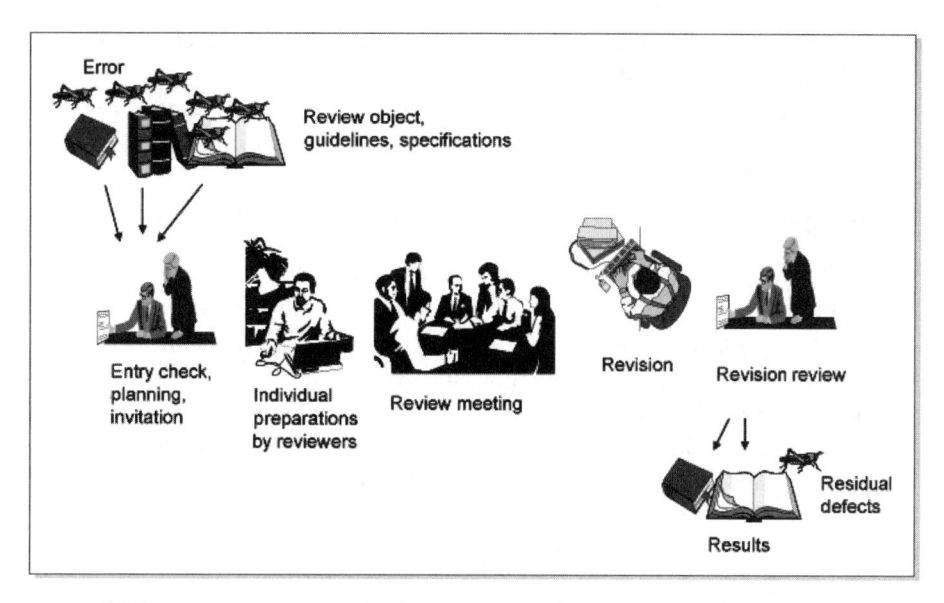

Figure 2-13 Example of a review process

BP3: Prepare joint review. Collect, plan, prepare and distribute review material as appropriate in preparation for the review.

NOTE: The following items may be addressed: Scope and purpose of the review; Products and problems to be reviewed; Entry and exit criteria; Meeting agenda; Roles and participants; Distribution list; Responsibilities; Resource and facility requirements; Used tools (checklists, scenario for perspective based reviews[75] etc.).

Review objects are collected by the review leader according to the review plan and distributed to the checkers. Depending on the requirements of the review process, the review leader performs at least sample checks (entry check) to see whether the review objects are of sufficient quality to perform the review. If this is not the case he may cancel or postpone the review to enable efficient review execution at a later stage.

BP4: Conduct joint reviews. Conduct joint management and technical reviews as planned. Record the review results.

Reviews are conducted according to the review plan and the results are documented (see review records, WP-ID 13-19). Besides general information such as

---

75. In perspective-based reviews the test object is checked from different perspectives. For this purpose, each checker is given a scenario that describes one perspective (e. g., from the test engineer's perspective).

date, review object, participants, and result of the reviews, detected defects in particular must be recorded.

Automotive SPICE focuses on the identification of problems. On a more general level, [IEEE 1028] talks about »anomalies«. Besides problems and defects, reviews frequently also identify open issues, ambiguities, and improvement proposals. These, too, should be documented and dealt with after the review.

In formal reviews, actual review execution typically consists of two steps:

1. Individual preparation of the checker: checkers prepare themselves individually and check the review object—with the aim to find as many defects as possible. Depending on the method and process used, review aids such as checklists and scenarios are used.
2. Review session: the review facilitator chairs the meeting and detected defects are discussed—with the objective to identify the defects unequivocally and comprehensibly so that the author can correct them afterwards. Detected defects are classified and prioritized as needed.

The theory of many review methods requires participants to confine themselves to problem identification during the review meeting. There is to be no discussion about solutions during the review session to avoid lengthy review meetings[76]. However, it is part of human nature that people tend to immediately discuss possible solutions when faced with important problems. It therefore helps if, after the review meeting, people agree on on a date for a problem solving session to deal with complex problems (e.g., a half-hour brainstorming session). After the review, the review results are defined. Possible results are, for instance:

- Review passed without conditions
- Review passed conditionally; if the identified defects are resolved by ...
- Review to be repeated

BP5: Distribute the results. *The review results shall be documented and distributed to all the affected parties.*

The review findings are communicated to all affected parties. They may include the following:

- Distribution of detailed review minutes according to the distribution list (at least to all participants; if required, also to the quality assurance group, project management, and teams affected by the revised review object[77])
- Distribution of a summary to management[78]

---

76. One empirically established figure is that review meetings should not exceed two hours in length as effectiveness and acceptance, for instance, participants' motivation, will decrease.
77. For example, the test team must be informed about changes resulting from the a requirements document review so that it can adapt test cases accordingly.

BP6: Determine actions for review results. The review results shall be distributed and analyzed; resolution(s) for the review results shall be proposed; priority for actions determined.

Distribution was already discussed in BP5. Defects to be eliminated were identified in the review (see BP4). Now the review object must be revised and the defects corrected. Normally this is done by the author, who will analyze the results and work out solutions. The classification of a problem already determines its priority; moreover, priorities can be explicitly set in the review session.

BP7: Track actions for review results. Track actions for resolution of identified problems in a review to closure.

After the defects have been removed (see BP6), a check must be performed to see if all identified problems are actually resolved. Depending on the rules set out in the review process, this is done by the review facilitator or the author himself.

BP8: Identify and record problems. Identify and record the problems detected during the reviews according to the established mechanism.

The recording of problems identified during reviews is already described in BP4 and BP5. Moreover, it may be worth considering recording problems at the organizational level for the purpose of process improvement.

2.15.4 Designated Work Products

13-19 Review records

Besides general information like date, review object, participants, result of the review[79], and effort, detected defects must be logged. A review log is suited to this purpose (see figure 2–14). Moreover, identified improvements proposals resulting from the review should also be documented. In case of less formal review methods, projects sometimes do without a review log. If, in this case, handwritten records of the review results exist (e.g., notes on a print-out of the review object), they should be kept as proof.

---

78. Depending on the culture prevailing in the organization, distribution to management can be problematic if results are used to conduct performance ratings that raise fear and anxiety among developers. As a consequence, problems are no longer admitted, review meetings are hampered by quarrels, and elaborate and expensive »private reviews« are conducted prior to official ones.
79. Possible results are, for instance, »review passed without conditions«, »review passed conditionally, if the identified problems are removed by ...« or »review must be repeated«.

Review Issue Log

Object:	
Version:	
Reviewer:	
Date:	

#	Document Reference	Where (page/ chapter/line)	Issue Severity	Issue Type	Checklist	Brief Description of Issue	Further Comments	Author evaluation
1								
2								
3								
4								
5								
6								
7								
8								
9								
10								
11								

Legend	Description
Document Reference	Document name
Where (page/chapter/line)	To which part of the document does the comment refer to
Issue Severity	Maj: Major Min: Minor
Issue Type	I: Improvement Proposal ?: Question to Author D: Defect
Checklist	Reference to checkpoint of checklist being used
Brief Description of Issue	What is it about?
Further Comments	Further commentary – Improvement proposals, change proposals, issues to be clarified
Author evaluation	Comment refering to comments/classifications resulting from the analysis of the issue

Figure 2–14 *Sample structure of a review log*

2.15.5 Characteristics of Level 2

On Performance Management

Review planning is done at two levels:

1. At project level: The number of reviews to be performed is defined on a per project basis. The reviews are then, for instance, included in the schedule of the respective projects. In large projects with many reviews a separate review plan may be drawn up.
2. Per review: Planned are the review objects, the participants, required resources, schedule, location, reference material, and suitable guidelines (review process, checklists).

On Work Product Management

The requirements of the Process Attribute PA 2.2 apply particularly for the review planning at project level and the review records. As explained in relation to BP7, after a review it should be verified if identified defects have actually been removed.

2.16 SUP.8 Configuration Management

2.16.1 Purpose

The purpose of the Configuration management process is to establish and maintain the integrity[80] of all the work products of a process or project and make them available to concerned parties.

In the context of configuration management (CM), the configuration management system is of central importance. By this we mean the combination of one or several CM tool(s) to support physical storage and handling, and associated rules, such as instructions, processes, and conventions; the latter, for instance, for change management, version control, or access restrictions. A CM tool does not necessarily need to consist of specific configuration management software. It may simply be a matter of storing files on a file system. The fewer the capabilities that a CM tool offers the more it must be supported by appropriate rules.[81]

2.16.2 Characteristics Particular to the Automotive Industry

Due to simultaneous development in different disciplines (hardware, mechanics, software, etc.) we are facing a heterogeneous tool environment. In software we typically find that one of the standard CM tools is deployed. Tools for mechanical construction (CAD) and circuit board layout usually have their own integrated version-control system. Bills of materials are, for instance, managed via SAP. Often certain documents (e.g., specifications, design documents) are deliberately kept outside the CM tool and stored on the file system, e.g., under Windows. The most common reason for this is that not everyone involved in the project has access to a CM tool or that they are not trained in the use of the tool.

This constellation makes configuration management more difficult than in a software-only project: the organization must create a system that allows, at certain points in time (e.g., prototype delivery dates), the identification of all configuration items (e.g., documents, code modules, data, development environment) relating to a defined level of development (a so-called baseline). This may be done by drawing a local baseline in each of the tools and specifying the associations in a table. Alternatively, data may be stored in the CM tool and then be automatically mirrored onto a generally accessible file storage medium. Most CM tools support this approach. In practice, configuration management in the automotive industry often shows weaknesses in the coordination of these disciplines.

---

80. In information technology, data integrity is understood to mean correctness and consistency of data. According to IEEE Std. 610.12-1990, integrity means the prevention of unauthorized access and unauthorized change of data.
81. Storing data on a file system there is, for instance, no automatic version control. Version control, in this case, is carried out manually according to strictly specified rules.

> **Note to Assessors**
>
> As a minimun, the following issues should be assessed in this process:
>
> - If there are different disciplines in the project (hardware, mechanics, software, etc.), is there a convincing concept for a CM spread across tools? Have mismatches between the development tools been compensated for?
> - Is the selection of the CM elements sufficient in principle to define a baseline? In other words, is it possible to reconstruct a baseline of a specific state of development based on CM elements?
> - Are changes traceably documented?
> - Are consistency checks of individual CM elements and baselines performed (at least prior to delivery) ?
>
> In any case, the CM tool should be looked into and sample checks should be done on changes to CM elements to examine if changes are comprehensively reflected in the change history.

2.16.3 Base Practices

BP1: Develop a configuration management strategy[82]. *Develop a configuration management strategy, including configuration management activities and a life cycle model, responsibilities and resources for performing these activities.*

NOTE: *The configuration management strategy should be documented in a configuration management plan.*

NOTE: *The configuration management strategy should also support the handling of product/software variants.*

This includes the basic configuration management activities and associated dates, such as:

- From when onwards particular CM items are to be put under CM control (e.g., from which milestone onwards or from which maturity level of the configuration items).
- The configuration management tools.
- The procedure for how to create baselines. This is particularly important in heterogeneous tool environments (compare with section 2.16.2); for instance, if a baseline correlating to a multi-tool environment is specified in one table.
- Procedure for how to manage product and software variants.
- Procedure for how to support integration using configuration management.
- Necessary steps to create product versions using configuration management tools.

---

82. NOTE: At Level 1, this base practice already requires basic planning elements whereas at Level 2 more elements are required by the generic practices of Process Attribute PA 2.1.

2.16 SUP.8 Configuration Management

- Procedure for how releases are carried out in connection with deliveries (i.e., whether the product contains the right components, whether the prescribed tests have been executed, who authorizes the delivery and in what way).

The results are typically documented in a CM plan.[83]

BP2: Identify configuration items. *Identify configuration items according to the Configuration management strategy that need to be stored, tested, reviewed, used, changed, delivered and/or maintained.*

NOTE: Items requiring configuration control should include the products that are delivered to the customer, designated internal work products, acquired products, tools and other items that are used in creating and describing these work products.

NOTE: Software development configuration items typically are e.g.,:

- Configuration management plan
- Requirements documents, architecture and design documents,
- Software development environment,
- Software development plan,
- Supplier agreements,
- Quality plan,
- Software units (code) including documentation,
- Test cases and test results, review documentation,
- Build list, integration reports and
- Customer manuals.

NOTE: Items requiring configuration control may also include work products of hardware (layouts, drawings, circuit boards, bills of material, etc.) and mechanical development.

CM elements are items (e.g., files) that are put under CM control. These are specified in the CM plan where, of course, only the file type is specied rather than each individual file (i.e., source code rather than code file xyz).

In most cases, only a selection of the work products is put under CM control. This selection must allow qualified baselines to be drawn (see BP5), i.e., ensure that in effect all elements (requirements and design documents, development environments, change requests, test cases, test documentation, also perhaps important intermediate work products, etc.) describing a particular development stage are available for baselining in the CM system. This way, individual development stages can be completely reproduced at any time at the »touch of a button«. Development stages include internal ones and those resulting in deliveries to the customer.

---

83. Normally, the CM plan includes additional planning elements which are not required in this BP but only at Level 2 (see also previous footnote).

In practice, a very common problem found in software projects is that only code is put under CM but none of the associated documents and files (see list in the previous paragraph). However, BP5 of the standard requires that such work products are baselined.

BP3: Establish a configuration management system. Establish a configuration management system, which provides an efficient means for handling the configuration items.

NOTE: *A configuration management system may include storage media, structures and hierarchies, procedures, access control and adequate tools for accessing the configuration items.*

As already described, a configuration management system (CM system) is of central importance to carry out efficient configuration management. In general, the less automation there is, the more effort must be spent by individuals in charge of CM, for instance, to check the correctness and completeness of baselines or to see that manual version control is properly done on a filing system. Other important aspects of a CM system are support in case of distributed development or of the interfacing of different CM systems between suppliers and OEMs. In distributed development, a jointly managed CM tool or at least a jointly managed data filing system would most certainly be of advantage. However, one thing which must not be forgotten in these scenarios is access protection. The following questions are particularly important with regard to a common CM system:

- Who is entitled to access CM elements, e.g., source code or project-planning documents?
- What is the role of the system administrator and what are his rights?
- Are speed and availability sufficient?
- How are joint activities synchronized (file locking mechanism, semaphores)?

In case of separate CM systems at least the following questions ought to be answered:

- What data is exchanged when?
- Where is the master for which CM elements?
- Are baselines spread across independent systems and how are they drawn and managed?
- How does exchanged information or how do files enter the local CM System?
- How do we make sure that changes have been consistently incorporated (e.g., in local copies) ?
- And the same question as for a common CM System: How are joint activities synchronized (file locking mechanism, semaphores)?

In both approaches, a branch management strategy (see BP4) may have to be considered.

BP4: Establish branch management strategy. *Develop a branch management strategy where applicable for parallel development efforts that use the same source base.*

NOTE: *A branch management strategy will include branch management, merging strategies, configuration item versioning in a branching system, branch parenting strategies and labeling strategies.*

NOTE: *A branch management strategy will define why and when branches will be created, which activities will occur in the branches, and how the branches will complete and/or migrate into the main source base.*

There are cases where one and the same CM item is subject to concurrent development, for instance, if several developers work on the same code component simultaneously. This happens frequently if developers work in parallel and correct defects. In these cases developers work on physically separate copies of the original configuration item, i.e., on a so-called »branch« of the master CM item. In the end, all changes must again be merged. The merging functionality, however, cannot prevent errors from slipping in as a result of parallel development. In such cases there are increased requirements on the quality assurance of the associated configuration item. For this reason, the strategy could prescribe that after successful merging a review needs to be performed.

Regulation is therefore needed as to whether and in which cases branching is permitted, how copies of the configuration items are to be named and versioned, how merging is to be done, which verification steps (e.g., reviews, tests) are required and what the interaction with baselines is supposed to be. Advanced CM tools in software development environments support branching through automatic version control and automatic merging, indicating conflicts should they arise.

BP5: Establish baselines. *Establish the internal and external (delivery) baselines according to the configuration management strategy.*

NOTE: *In complex software systems with many work products, the preparation of external (delivery) baselines may be supported by the reasonable use of several intermediate internal baselines.*[84]

NOTE: *For baseline issues refer also to the product release process (SPL.2).*

A baseline denotes a grouping of CM items identifying a specific state in development. Associated configuration items are identified accordingly. A particular, non-modifiable version is taken of each configuration item. Only later versions of

---

84. If a project is made up of several subprojects, separate baselines are often drawn up for individual system parts. If for example a version of the overall product is to be created, it is necessary to establish an »overall« baseline. Subbaselines may exist in subsystems, other disciplines (hardware, software, etc.), other CM tools, ...

it can be created. This way, items are protected against changes and the state of development can be reconstructed at any time, either in part or as a whole. Items belonging to a baseline must be consistent with each other (see also BP10).

Baselines should be drawn whenever there is a delivery (i.e., release). In addition, baselines should also be created for internal purposes, for instance, if requirements are stable, if the design is mature, or if a release for tests or trials is being produced.

To uniquely identify baselines, many CM tools provide a mechanism to »label« baselines, i.e., assigning an identical label to all versions of the affected CM items, for example »3rd delivery status«. A distinction is made between »floating« and »fixed« labels. A floating label is always attached to the latest version of a CM item, i.e., it automatically shifts to the latest version. Thus, it is very easy to identify the latest state of development. In contrast, a fixed label cannot be shifted. With this method CM items can be »frozen« for the purpose of baselining.

BP6: Maintain configuration item description. *Maintain an up-to-date description of each configuration item.*

NOTE: *The description should identify:*

- *its decomposition into lower level configuration components;*[85]
- *who is responsible for each item; and*
- *when it is placed under configuration management.*

Such descriptions are usually stored in different places, for instance, in the CM tool or in the document itself. Examples are:

- Description/elaboration of a CM item, including classification into a possibly existing CM items hierachy
- Description of a modification to a configuration item (e.g., in the CM tool prior to check-in) or in the configuration item itself (in a change history)
- Responsibility for a CM item, author, etc.

Regardless of individual CM items, item descriptions may also be stored in project descriptions or in the CM plan, e.g.,:

- Description of a CM item type
- Definition from when onwards a work product type is to be put under CM (e.g., from a particular development phase or from a particular maturity level of the work product).

Rules must be established for these descriptions (e.g., in the CM plan or in process descriptions), for example:

---

85. This means decomposition into smaller CM items of a lower aggregation level.

2.16 SUP.8 Configuration Management

- How to keep the change history in documents.
- How changes in the CM system are to be documented.
- What the versioning strategy will be (especially if the CM tool does not provide this functionality).

BP7: Control modifications and releases.[86] *Establish mechanisms in order to determine configuration items to change, check in/out, configuration item access permissions, version identification and change, change commenting, configuration item locking/commit.*

It must be possible to flag a CM item to be changed, for instance, by setting its status to »In revision«; also, to check items in and out of the CM tool. If a CM item is checked out, it should be either locked for other users, i.e., only allow read access, or require the creation of a branch (see BP4) before it is checked out. During check-in of a changed CM item, a new version number must be assigned and the CM element is released again. Unpermitted access must be prevented. Changes compared to a previous state must be described (usually in both, the CM system and in the CM item). All actions are logged by the system.

In many CM tools (e.g., for software, CAD, printed circuit board layout) these mechanisms are preset by the tool. However, this does not mean that these facilites are actually used by staff (for instance, have relevant comments been provided during check in?). A bigger problem arises with work products that are stored in the file system, because there the mechanisms must be carried out manually (unique versioning in file names, change history in the document, access rights, etc.). Arrangements for appropriate ruling for manual configuaration management must, for instance, be made in the CM plan.

BP8: Maintain configuration item history. Maintain a history of each configuration item in sufficient detail to recover a previously baselined version when required.

What is meant is the requirement for a history of the changes to a CM item. There are two places where to maintain the change history:

- In the CM tool
- In the work product

What is best depends on the individidual case and varies from work product to work product. If the work product is kept in the CM tool, its functions should definitely be used for the documentation of changes. If this is not the case, the work product itself should include a change history. Individual modification

---

86. Here releases are understood in the sense of product releases with corresponding baselines, i. e., as CM items, too. Planning (dates, contents) is done as part of project planning. Further practices are described in SPL.2 Product release.

notes must however be detailed enough to be able to reconstruct the changes. Especially important in the context of change management is that change-IDs are recorded (see SUP.10).

Common weak points are missing or meaningless modification notes or even completely missing control and change documentation. The latter can frequently be found in work products that have been modified several times (e.g., schedules), often only one single version exists (the current one), and changes have not been documented.

BP9: Report configuration status. Report status of each configuration item.

NOTE: *Regular reporting of the configuration status (e.g., how many configuration items are currently under work, checked in, tested, released, etc.) supports project management activities and dedicated project phases like software integration.*

The development status of CM items is to be reported to support project tracking. Typical states are buggy, locked, in work, reviewed, tested, approved, etc. This base practice is all the more important the larger and more complex a project is. It is particularly important for the following activities:

- Integration activities: The integration representative obtains an overview about the work status of the individual developers. A CM tool provides this information at the touch of a button.
- Change management activities: Which change requests are still outstanding, e.g., for a particular release? If a change management tool is used that is interfaced with a CM tool, the individual states of change requests (e.g., requested, under analysis, in process, in test) can usually be provided at the touch of a button.
- Project management activities: What is the current project status?

BP10: Verify the information about configured items. Verify that the information about configured items, their structures and baselines, supplied through status accounting reporting is complete and ensure the consistency of the items and baselines.

Checks must be performed and resulting corrective measures must be taken to ensure that CM items, baselines and the CM system are consistent. A vital question to ask is how to ensure that a baseline was drawn appropriately (especially if we talk about a release version to be delivered to the customer). Are the correct files included (this is particularly relevant in case of parallel development, variants, distributed development or gaps in the tool chain)? Have all the changes planned for the delivery been actually included? Was there a check to see if the CM items were actually tested according to plan? This kind of check is also known under the term »CM Audit« or »Baseline Audit«.

2.16 SUP.8 Configuration Management

A CM Audit is defined as a check of the *entire* CM system and its structures, sample checks of conventions, etc. The results of these checks must be reported to the relevant project staff (e.g., the CM representative and the project manager).

Insufficient verification measures can be recognized by the fact that already corrected defects re-emerge from time to time or that committed functions turn out not to be included in the release version[87] (and that no notification of this is given).

BP11: Manage the backup, storage, archiving, handling and delivery of configuration items. Ensure the integrity and consistency of configuration items through appropriate scheduling and resourcing of backup, storage and archiving. Control the handling and delivery of configuration items.

Important questions include:

- Is it guaranteed that backups (regularly, e.g., weekly) are drawn and archiving is done (after project completion)? (One example of a typical weakness would be: Backups are made regularly but based on a rolling system of tapes. Each week, for instance, a backup tape is created. There are, however, only 20 tapes and after week 20, tape 1 is overwritten. Since the project lifecycle is a lot longer, the recovery of older versions, if required, turns out to be impossible.)
- Are all work products delivered to the customer (e.g., software, hardware, development documentation, test logs) consistent with each other? Does the software identify itself in accordance with the declared version?
- During a product delivery comprising software and hardware: How are software and hardware put together? Which software version matches which hardware version? Who will load the software onto the hardware, and how? Which final checks are done? How does one, for instance, ensure that version information is consistent (sticker on the hardware, software ID, version according to accompanying documentation, etc.)?

There must be sufficient resources (i.e., staff and technical infrastructure) available to handle all these issues. Associated activities must be planned and scheduled.

2.16.4 Designated Work Products

01-00 Configuration item

Configuration items (CM items) identify objects which are managed using configuration management. These can be software-related items like software units, subsystems, and libraries, but also test cases, compiler, data, documentation,

---

87. Such symptoms are a certain indication that other, additional weaknesses exist in the CM system.

physical media, and external interfaces. A least version identification should be maintained and a description should be available specifying, for instance, type, person responsible, status information, and relationships to other CM items. In most cases interdependencies exist between CM items, for instance, between software units of one subsystem to software units of another one. These dependencies facilitate integration, defect removal, changes, and support traceability.

08-04 Configuration management plan

The CM plan is the central CM planning instrument. It contains all relevant stipulations which are important for the project from a CM perspective. Concrete scheduling for CM activities is usually found in other planning documents (i.e., in the project schedule). Many organizations provide a generic CM plan from which the project-specific CM plan is derived.

Example structure of a CM plan
1. Roles and authorization regarding CM
2. Involved staff
3. Project structure from the CM perspective
4. Product structure from the CM perspective
5. CM items
    - 5.1 CM item types
    - 5.2 Naming conventions for CM items
    - 5.3 Development tools under CM
6. Baselines and Releases
    - 6.1 Naming conventions for baselines and releases
    - 6.2 Planned baselines and releases
7. The Build process
    - 7.1 Reproduction of a configuration
    - 7.2 Description of the build process
8. CM environment
    - 8.1 CM tool(s)
    - 8.2 Physical storage location(s)
    - 8.3 Structures of the CM library(-ies) and interaction
    - 8.4 Access rights
    - 8.5 Backup of the CM library(-ies)
    - 8.6 Consistency checks
9. Change Management
10. Archiving of releases

16-03 Configuration management library

There are two different points of view concerning what constitutes a configuration management library (CM library). It is typically understood to be the central storage of the project-specific CM system from which at any time the correct products, including release and test configurations, can be recovered, and by means of which the status of the CM items can be identified. However, the CM library can also serve as a tool for reuse across projects and in that case includes a library of reusable software units, function descriptions, test cases, etc.

2.16.5 Characteristics of Level 2

On Performance Management

According to Automotive SPICE, CM planning takes place in two steps: At Level 1, several base practices, e.g., BP1, BP2, BP4, and BP11 already require basic planning items, that is to say the principal procedure, provision of resources, and a schedule for CM activities. At Level 2, more planning requirements are added with the generic practices of Process Attribute PA 2.1, e.g., regarding the assignment of responsibilities and the planning of resources (see section 3.3.3, deliberations to PA 2.1).

Regarding the problem of scheduling: for many CM activities (for instance, the frequent check-in and check-out of configuration items), scheduling does not make sense. However, reasonably plannable are activities related to the creation of releases or CM audits.

On Work Product Management

GP 2.2.3 requires features of configuration management for the configuration management process itself. While the process of configuration management is dedicated to the work products of a project as a whole, GP 2.2.3 aims at the work products of the process. The latter, for instance, means that the CM plan itself, being a central work product of the process, needs to be put under CM control, i.e., that it must at least be versioned and provide a change history.

2.17 SUP.9 Problem Resolution Management

2.17.1 Purpose

The purpose of the Problem resolution management process is to ensure that all discovered problems are identified, analyzed, managed and controlled to resolution.

Problem resolution management is about the ability of an organization to identify, analyze, monitor, control, and remove problems in a structured, traceable way. In practice, implementation of this process is often based on a status model. Such a model contains the different states of a problem (e.g., accepted, in progress, in analysis, see figure 2–15) as well as the conditions for state transitions. Collecting statistical data, for example about the duration of a problem in a particular status and the analysis of data in the form of a trend analysis helps to determine improvement or deterioration in a project's problem management over time, derive trends and, if necessary, initiate counter measures. Problem resolution management can be applied to problems and defects of all kinds and is a major factor contributing to customer satisfaction. This process is closely connected to the change management process (SUP.10). SUP.10 describes the work flow within change management for all types of changes, not only for problems. As far as the management system is concerned, problems and other changes (e.g., requirements changes, design changes) are managed, processed, and tracked in the same way. SUP.9 focuses on the work steps especially required for problem resolution. With BP7 it has a clear interface to SUP.10: a diagnosed defect is the basis for changes of work products and is further processed within the change management system.

> **Note to Assessors**
>
> Due to similar process requirements, identical treatment, and identical responsibility, interviews of both processes are frequently combined. Attention should be paid to the interaction of the two processes: SUP.9 provides possible input for SUP.10.

2.17.2 Characteristics Particular to the Automotive Industry

In the development of vehicle systems there is a start of production (SOP) date, at which point the product (a system consisting of hardware and software) is supposed to be 100% completed and fault free. Each defect detected later will lead to repairs or recalls and associated high costs, as well as to an image loss for the manufacturer. Compared with the development of IT systems (where defect removal is easily possible within the context of successive software releases), a fully functional problem resolution management is of increased importance.

Complex vehicle components consist of many subsystems with distributed functions. A problem which occured at a particular location in the system need not necessarily have been caused at this location. This makes localization of problems more difficult.

2.17.3 Base Practices

BP1: Develop a problem resolution management strategy. Develop a problem resolution management strategy, including problem resolution management activities and a life cycle model, responsibilities and resources for performing these activities.

2.17 SUP.9 Problem Resolution Management

Minimum elements of this strategy are:

- Specifications regarding problem handling, including required activities (problem description and problem capturing, data collection, analysis, correction, etc.)
- Staff with assigned responsibility for performing these activities
- A system for problem tracking (a defined lifecycle including problem status, and problem prioritization)
- Predefined time periods for problem handling (maximum time for a state transition)
- Guidelines regarding feedback to involved staff (the person reporting a defect, project team member(s), management)

The problem resolution management strategy may be part of a problem management plan (Level 2) and should be able to interact with the change management strategy (SUP.10, BP1).

BP2: Establish a consistent problem resolution management proceeding. *A problem resolution management proceeding is established in order to ensure that problems are detected, described, recorded, analyzed, resolved and prevented in a consistent and traceable way based on the problem resolution management strategy. Interfaces to affected parties are defined and maintained.*

Based on the problem resolution management strategy, a suitable method must be specified and documented (e.g., in the form of a problem management plan). It typically begins with the reporting and logging of a problem and ends with its final and verified removal. It is necessary to channel the paths a problem report takes. To do so, a central contact person should be assigned and this should be documented in the project together with all the other responsibilities. It is good to have appropriate tool support using, for instance, change management systems or databases that are available via the intranet or Internet.

In larger projects with subprojects it is often difficult to keep track of open problems and their status at particular deadlines (e.g., prior to deliveries). In order to avoid elaborate manual collection of information, it is advisable to establish a problem resolution management system that fulfills the following requirements:

- Information related to problem description is recorded as accurately as possible. Besides a description of the problem this also includes the conditions under which the problem arose, date of occurrence, reporting person, etc.
- Each problem is uniquely identified, e.g., by means of a numbering scheme.
- Each problem is assigned to a person for analysis, resolution, and so on. Responsibilities regarding resulting work steps are clearly assigned.
- Different problem states are distinguished. This way it is possible at any given time to determine a problem's exact state and the time it has remained in that

state. If the lag time in a status exceeds the allowed period this can be noticed and reacted to quickly. For example, it may have been defined that a customer is to receive information about the solution of a problem together with a brief description of its cause and date of correction not later than three days after the problem was reported. Figure 2–15 illustrates that after a problem or defect is detected, its status is generally set to status »open«. Vital for further treatment is the subsequent problem or defect analysis (status »in analysis«). Based on the results, further steps are initiated that will finally lead to the resolution of the problem. The status is then set to »closed«.

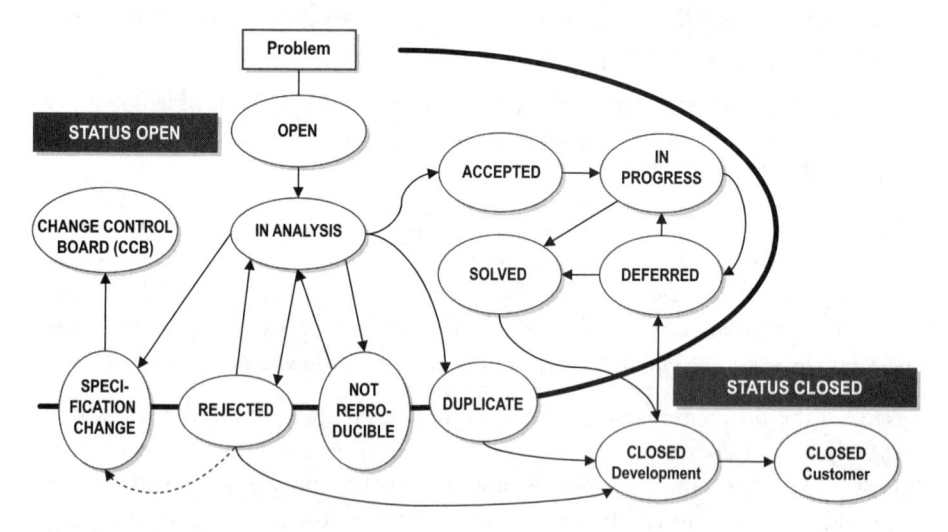

Figure 2–15   *Example of a problem status*

- Priorities for problem treatment are assigned (see on this point also SUP.9 BP4).
- Scheduling is done (e.g., defining a due-date for problem resolution). Customer, end customer, and problem reporter respectively expect feedback related to their problem. This is the reason why it is defined when and how often a problem reporter is to receive feedback regarding the progress of problem solution. The following customer-oriented time management method has proved successful in practice:
    - The problem reporter receives immediate feedback that his report has been recorded, including a reference number (ID) for possible queries,
    - Feedback about the expected correction date (as soon as known) and
    - Feedback about successful closure including indication of impacts or changes.

2.17 SUP.9 Problem Resolution Management

BP3: Identify and record the problem. Each problem is uniquely identified and recorded.

NOTE: *Problems are typically recorded in a database. The necessary supporting information needs to be provided in order to support diagnosis of the problem.*

NOTE: *Unique identification supports traceability to changes made.*

A precondition for the solution of a problem is its description. The more concrete and more detailed the description, the easier subsequent root cause analysis will be. If only for the sake of traceability and later detectability, problems must be uniquely identified (e.g., by means of an ID) and centrally managed after recording (e.g., by means of a database). Problems[88] and other changes are often jointly managed. There is no objection to this if the later division of the two entities (e.g., for a statistical evaluation) is possible. Moreover, not every recognized problem will necessarily lead to a change.

Recording and administration of accepted problems is done in a problem resolution management system or in lists that include all necessary information of a problem record, for instance:

- Problem number (ID)
- Priority
- Date (of the report)
- Impacted hardware version, impacted software version
- Problem reporter
- Problem description including constraints and all available additional information that supports problem diagnosis
- Analysis result
- Impact of the problem on other systems/areas
- Necessary measures and changes, including unique change ID
- Person responsible (for resolution)
- Target release version
- Due-date

Figure 2–16 shows an example of a problem or change report.

---

88. A problem is not a defect. A problem may have many causes, a defect is one of them. A problem does not necessarily lead to a change.

Problem or Change Report			
Project: 210815			**Problem / Change ID:** 018
Reporting Person: Steve Briggs (Starks & Brothers)			**Priority:** A
Problem Report: ( ) **Change Request:** (X)		**Date:** **Date:** 12.17.07	**Affected Hardware Version:** **Affected Software Version:** 2311
Description: Revision and updating of the auditing tool according to the latest ISO 9001 version			
Description of the Problem / Change: Mr. J. Marley requests revision of the auditing tool, to be followed by a review meeting (2–3 h) with the entire QA Team concerning this issue. Tool changes will be discussed and results will be logged online. Any further changes resulting from this meeting will also be incorporated in the auditing tool. Mr. P. Seeger is to assume the role of the DIN/ISO 9001 expert. Clarification needed as to whether additional activities (preparation, follow-up documentation, meetings) are necessary. If yes, approx. 2 h will be needed for each additional meeting.			
Objective / Rationale: Results and contents are necessary for audits to comply with the standards.			
Analysis results:			
Consequence if change is not implemented: No further audits.			
Measures to be initiated: Tool to be revised.			
Person in charge: Jack Manning			
Estimated implementation effort: 2 meetings at 3 h each (including preparation = 6 h, if preparation and follow-up by Starks & Brothers + additional 10 h)			**Target version:** 2911 **Date:** 03.24.2008
Costs:			**Benefit:** Availability of revised, enhanced tool and documentation.
Project internal consequences (including list of documents/work results to be changed): none			**Project-external consequences:** none
Release by steering committee (if applicable, reference to steering committee decision): not necessary			
Technical release project manager Phil O'Connor, QA **(Date/signed):**			**Technical release project manager Steve Briggs** **(Date/signed):**

Figure 2–16   Example of a problem or change report

2.17 SUP.9 Problem Resolution Management

BP4: Investigate and diagnose the cause and the impact of the problem. Investigate and diagnose the cause and the impact of the problem in order to determine appropriate actions and provide classification.

NOTE: *Problem classification (e.g., A, B, C, light, medium, severe) may be based on severity, impact, criticality, urgency, relevance, etc.*

Problems are classified upon receipt (see also figure 2–17). At the same time the problem reporter is informed about the receipt. A precondition is that all information necessary for problem resolution and problem classification is fully available or can be obtained at short notice (i.e., through queries to the problem reporter). Problem classification serves to prioritize the problems and to derive a sequence of steps for resolution. Criteria for prioritization are, for instance, criticality, urgency, impacts expected to the prime contractor or end customer.

Complaint category	A-Complaint			B-Complaint			C-Complaint		
Complaint rating	> 100	90	70	60	50	40	30	20	10
Complaint evaluation	Safety risk Unsellable vehicle Break downs	Not acceptable, Will definitely lead to customer complaint	Very serious complaint about the paintwork	Unpleasant, Bothersome, Complaint to be expected from average customer, Quality defect present			In need of improvement Complaint to be expected from demanding customer Demand on quality not met	If complaints accumulate, complaint to be expected from demanding customer	
Impact on customer	Vehicle not available	Vehicle must pay unscheduled visit to the repair shop		Customer will have complaint sorted out during next planned visit to the repair shop			In general, not all customers expect correction of the complaint	Customer complains about quality level	
Detectable	by all customers								
	by average customer								
	by demanding customer and trained auditors, taking into account internal quality standards								
Removal of complaint / problem	Complaint must be corrected, Ensure that no vehicle with this defect is delivered to the customer								
Corrective measures in series production	Initiating measures to remove cause so that problem does not reoccur						In case complaints accumulate, initiate corrective measures	Monitor, avoid deterioration	

Figure 2–17   Example of a problem classification scheme

Above all, particularly urgent measures (see BP5) must be recognized. Usually an analysis is carried out before any action is taken regarding problem removal. The reason for this is that usually only symptoms or impacts of a problem are reported that allow no direct conclusion about its cause. In practice, the analysis of a problem is typically assigned to the developer responsible for the assumed defect. He is responsible for detailed investigation and analysis, documentation, status updating, and—not least—for resolution within schedule. This does not necessarily mean that he needs to be doing all of these activities himself.

Many problems have a local dimension and can be eliminated with little effort. Some problems, however, require repair activities on different parts of the overall system before they are removed. Impacts thus identified provide important information for regression testing. Some problems have an impact that goes far beyond project boundaries: for instance, if a defect has been detected in a widely used software unit (e.g., drivers, platform software), other projects or even products that may have already been delivered to the customer are usually affected, too. In these cases much more comprehensive measures will be necessary.

As a supplementary measure, a »root cause analysis« (RCA) could be performed for process improvement purposes. In an RCA, not only the problem but also its causes (e.g., process gap, tool problem) are analyzed and removed. This is the only way to prevent future recurrence of the problem.

BP5: Execute urgent resolution action, where necessary. *If the problem warrants immediate resolution pending an actual change then obtain authorization for immediate fix.*

NOTE: *Following urgent resolution action, a problem resolution management proceeding should be established according to BP2 and progressed according to BP3, BP4, BP7 and BP8.*

In the case of severe problems with serious impact (e.g., severe malfunction, system failure, the system cannot be started-up, or if there is danger to life and limb) is it necessary to take immediate action. At the very best, contingency plans are already drawn up that can be implemented without delay.

Possible immediate measures are, for instance, the immediate removal of the defect cause, perhaps in connection with the shut-down of already running systems. Possible immediate, external measures may include issuing warning notices or recall actions. Such measures require authorization by management. Most of the immediate measures affect the entire project, i.e., supplier and OEM.

In order for a project or organization to be able to cope with the implementation of immediate measures, necessary rules and conditions (contingency plans) must be established beforehand as part of the problem resolution management strategy. As a result, the following parameters may be defined:

- Pre-defined organizational alarm levels
- Clear division of responsibilities and tasks between supplier and OEM in case of emergency
- Decision-making authority or, alternatively, escalation mechnism depending on the alarm level
- Further course of action, e.g.,:
    - A team of experts is formed for problem clarification.
    - Clarification is addressed with highest urgency.

- A package of measures or perhaps several options are worked out and agreed on according to the decision making rules.
- Effectiveness of measures is verified.
- Subsequent to immediate measures initiated, a detailed analysis is performed and causes of problems are eliminated to provide lasting and accurate defect removal (see note).

BP6: Raise alert notifications, where necessary. *If the problem is of high classification and impacts other systems or users an alert notification may need to be raised, pending a fix or change.*

NOTE: *Following the raising of an alert notification, a problem resolution management proceeding should be established according to BP2 and progressed according to BP3, BP4, BP7 and BP8.*

Customer, user, end customer, other projects and other affected parties must be notified immediately. This applies particulary if the product contains severe defects or if using it represents a serious danger (see also BP5). This way any possible damage is supposed to be prevented. The corresponding course of action is documented in the problem resolution management strategy.

BP7: Initiate change request. *Initiate a change request for diagnosed problems.*

NOTE: *The implementation of the change request is done in the SUP.10 Change request management process.*

If a problem can be attributed to a defect and if a decision for its removal has been taken[89], actions for its removal must be initiated. Defect removal is done within the context of the SUP.10 process (if this is implemented) and is passed on to it in the form of one or several change requests. Changes are typically made in a whole range of work products (e.g., code, design documents, test documents), perhaps even in processes and work flows.

BP8: Track problems to closure. *Track the status of all reported problems to closure. Before closure, a formal acceptance has to be authorized.*

Careful record must be taken of problems and problem status. Problems that have remained for too long in a particular status (e.g., analysis taking too long) must be spotted, problems must not get lost or forgotten, and it must be guaranteed that committed completion dates are kept (for instance, allocation to a certain release). It is of help if a person or group (project manager, change manager, change request board (see SUP.10 BP6)) has supervisory capacity and if statuses are regularly discussed (e.g., during project meetings). Appropriate tool support is also very useful, e.g., in the form of a change management system with an

---

89. Under certain circumstances, to avoid any risk of further defects being injected in the context of defect removal, minor defects may not be removed.

interface to the configuration management system. Such systems often have a workflow component which can issue assignment to individual people and manage status transitions automatically. In most cases they also support statistical status evaluation so that current status and problem resolution can be tracked over time using diagrams. Figure 2–18 provides an example.

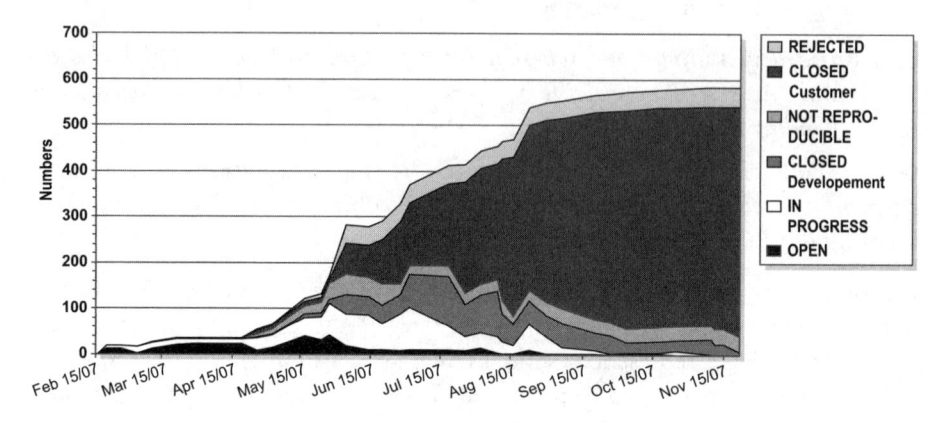

Figure 2–18   Example of problem and resolution tracking over time

Before a problem is finally closed, formal approval by authorized staff should be obtained (see also SUP.10 BP.11 on this point). The review result is documented.[90]

BP9: Analyze problem trends. *Collect and analyze data from the problem management system (occurrence, detection, affected range, etc.), identify trends and initiate actions, where necessary.*

It is good practice to collect data on problem status and solution progress at different levels, for instance, at product, process, project, and organization level, in order to identify trends related to problems and defects.

Individuals and organizations learn from problems and failures. The systematic elimination of defect sources opens up possibilities for cost reduction and quality increase. This can result in changed and enhanced processes, better tool support and improved guidelines, or staff specially trained in problem handling. The SUP.9 process provides an excellent data source which can be evaluated in the context of the process improvement process (PIM.3). This applies especially if statistical evaluations are available regarding defect categories, defect injection times, and affected product parts.

---

90. This may be important for legal reasons (product liability).

2.17.4 Designated Work Products

08-27 Problem Management Plan

The problem management plan defines and describes problem resolution activities from identification, recording, description, and classification right up to resolution. It also contains requirements and criteria for the evaluation and removal of problems. Furthermore, it describes tracking mechanisms for problem resolution activities and the collection and distribution of solutions and approaches. In practice these things are often regulated in a process description at the organizational level and possibly supplemented by project-specific commitments.

13-07 Problem Record

The problem record contains all the descriptive and relevant information pertaining to a problem. As far as an individual problem is concerned, the problem record contains the following:

- Author of the problem record
- Problem description
- Problem classification (e.g., criticality, urgency)
- Target due-date for resolution
- Measures for the verification of successful resolution

In practice, different names are used for the term *problem record* (e.g., test incident report, trouble ticket, etc.). Often a problem can be recorded using an available change management tool. The problem record serves as input information for a subsequent change and is therefore an important input for the SUP.10 process. What is important is that current status and assignment to a responsible person are visible. This work product is closely related to the change request WP-ID 13-16. In practice, the document template used for problem records and change requests is often the same. BP3 describes what belongs in a problem description.

15-01 Analysis report or
15-05 Evaluation report

The analysis or evaluation report documents the result of the problem analysis or corresponding evaluation. What matters is to know who carried out the analysis or evaluation at what time and with which result. The result of the analysis or evaluation must be adequately documented. The evaluation report is often part of the problem record and is kept there in a separate section (if applicable, of the database).

15-12 Problem Status Report

A problem status report contains information on current problems and their status in summary. This form of reporting is excellently suited to provide statistical evaluations on defect classifications, number of problems in different status, problem frequency, speed of resolution, lag time in a certain status, customer feedback, and so on.

2.17.5 Characteristics of Level 2

The subsequent deliberations also refer to SUP.10 Change management.

On Performance Management

Problem resolution management and change management are both particularly suited to prove management and control of processes and work products, and to project this externally. Since the external customer is behind many of the problem reports and change notes, it gives the development organization the opportunity to show that customer requests are considered an important part of their ongoing activities and that these requests are in good hands and systematically brought to completion. In practice, tracking of defects/problems and changes by means of metrics have proved useful (for instance, the identification of the lag time of a problem in a particular status), although at Level 2 this is not explicitly required in Automotive SPICE.

Individual processing steps are not planned at project level but planned and tracked by the individual person to whom the task has been assigned. This person is solely responsible for the activities assigned to him. At project level, project meetings are scheduled (during which problem management issues are discussed).

The completion status of the respective issues (problems/defects and changes) is regularly tracked (for instance, in weekly meetings). If it is noticed that set targets cannot be reached, counter measures that can be sufficiently evidenced are initiated. For instance, if problem removals planned for a particular delivery cannot be kept, additional resources must be provided for the task. This may be accomplished by outsourcing to an external service provider.

If changes occur, schedules are adapted accordingly. The status of problem removal (e.g., number of defects in the project) is regularly distributed and communicated to affected individuals, also—and particularly—to the external customer.

On Work Product Management

The requirements of Process Attribute PA 2.2, for example, specify the layout of a problem report, which fields of a problem input mask must be completed in which way, which requirements documents must satisfy, and how this and other relevant information is going to be communicated. This way a system is established which allows individual members of the project team easy access to information needed for their daily work (e.g., problem database or LOI[91] database). Work products are controlled by regular discussion and assessment of their status (e.g., in team meetings). Proof that it is done can be provided. The interdependencies between individual problems are identified and may determine the sequence in which they are solved. For instance, this is true for a problem report which leads to changes of the functional requirements, the architecture, the software design, and so on. Related information and documents are clearly assigned to a problem. Reported problems (e.g., problem lists) are regularly reviewed.

2.18 SUP.10 Change Request Management

2.18.1 Purpose

The purpose of the Change request management process is to ensure that change requests are managed, tracked and controlled.

The purpose of change management is to collect change requests in a structured and traceable way and to analyze, implement, and monitor their state of progress. The ability to do so is extremely important for an organization because there is no project without changes. On the contrary, many projects are literally inundated with changes. The primary reasons for this are:

- Changes caused by defect removal (see also SUP.9)
- Rapidly reduced product lifecycles, because manufactureres are forced to bring innovations onto the market in increasingly shorter cycles. The resulting drastic reduction of development lead times makes it increasingly difficult to exercise sufficient care during the early development phases (requirements, design). On top of that, manufacturers consider it necessary to respond to attributes of new rival products with the consequence that they allow many—even late—changes.

Many development organizations are unable to stem the tide of changes. The ability of an organization to control change management and still produce high quality products within the constraints of cost and schedule is becoming an increasing factor that differentiates competitors. In practice, there are two different levels of change management:

- The level that covers the project's contractual scope of work. At this level, small changes and bugfixes are managed which are not cost-relevant but are included in the development scope (level 1).
- The level that covers work which is outside the project's contractual scope: At this level, changes that are generally cost-relevant and not included in the development scope are negotiated, followed up, and implemented (level 2).

---

91. List of open issues.

Both levels differ by virtue of their decision-making power: At level 1, decisions are made within the project team, at level 2 higher decision levels, i.e., senior management, are brought in. At level 2, additional work steps are conducted, e.g., the issuing of a supplemental offer. The processes of both levels interact, i.e., a change approved at level 2 triggers implementation steps at level 1 (Base Practices 9-12). SUP.10 focuses primarily on level 1.

In practice, supplier and customer usually have already established their own, separate change request management systems at the organization level. When a change request management system is set up within a project, care needs to be taken that the existing change management systems can actually be linked with each other and that the requirements of all the involved parties on a common change request management system are equally considered.

The change request management work flow is very similar to the one used for problem resolution management (see SUP.9, also on the mutual demarcation of the two processes).

> **Note to Assessors**
>
> Due to similar process requirements, identical treatment and identical responsibility, interviews of both processes are frequently combined. (see also SUP.9). Attention should be paid to the interaction of the two processes: SUP.9 may provide input for SUP.10.

2.18.2 Characteristics Particular to the Automotive Industry

In the automotive industry, additional reasons may lead to requests for changes. In the majority of cases, reduced development times are only made possible by the massive deployment of simultaneous engineering, i.e., different product components are simultaneously developed but have so many cross-relationships that numerous changes become necessary during the development lifecycle. A different approach that is being pursued are platform strategies, i.e., platforms are developed which are independent from a model range. For each model range and every customer, custom-made variants of the general platform components are developed. If changes are needed or if defects occur with respect to the model independent components, all the development branches that build on them also need to be changed or adapted.

2.18.3 Base Practices

BP1: Develop a change request management strategy. *Develop a change request management strategy, including change request management activities and a life cycle model, responsibilities and resources for performing these activities.*

This strategy serves as a guideline and instruction for the recording, management, and processing of change requests. Change request management starts with the change request and ends with the evidenced implementation or rejection of the change. Change procedures and mechanisms are typically described in a change management plan.

Besides the points already stated in SUP.9 BP1 regarding a strategy, the following points need also be considered when working on a change request:

- Each change is subject to a feasibility analysis, possible risks (e.g., of a technical nature) are analyzed. The results are traceably documented.
- In practice, the release in which the change request is going to be implemented should be specified.

BP2: Establish a consistent change request management proceeding. *A change request management proceeding is established and implemented in order to ensure that changes are detected, described, recorded, analyzed and managed in a consistent and traceable way based on the change request management strategy. Interfaces to affected parties are defined and maintained.*

Especially in larger projects with subprojects it is necessary that one can obtain an overview of all the change requests and their implementation status at any time. In order to avoid laborious manual collection of information one is better advised to establish a change request management system. Moreover, it is also a good idea to determine a central contact person (e.g., the project manager) to be able to record change requests centrally.

In practice, change requests and problems are often jointly managed (on demarcation, see SUP.9). Occasionally, however, several change request management systems are implemented that depend on the type of change (e.g., requirements changes, other changes). This often happens if different groups are involved or if a project is very large. Usually the same or very similar guidelines are applied for the implementation of change requests and the resolution of problems (see also our deliberations in SUP.9 BP2).

BP3: Identify and record the change request. *Each change request is uniquely identified and recorded and the initiator of the change request is retained.*

NOTE: *Provide traceability to originating problem or error reports. Change requests submitted as a resolution to a problem or error report should retain a link to the originating problem or error report.*

The implementation of a change begins with its description. The more concrete and more detailed the description, the easier it is to perform the necessary analysis regarding feasibility and impacts, and to update associated documentation (e.g., requirements, architecture, design, test). Changes must be uniquely identified and centrally managed (for instance, by means of lists or a database) after they have been recorded. Often changes and problems are jointly managed. There is no objection to this if both can later easily be separated, for instance, to enable statistical evaluations.

Desired changes are recorded and managed in the change request management system or in lists which include all the necessary information (see also SUP.9 BP3). When a change request is recorded, the following points need also be considered:

- Results of the internal evaluation (feasibility analysis/risks)
- Reference to the problem or incident report it is based on

References to the underlying problem or incident report establishes the link between SUP.9 and SUP.10. On the one hand this is information for the person implementing the change request, on the other hand it is important for the tracking of the implementation status within SUP.9.

BP4: Record the status of change requests. *Change Requests and changes are allocated a status indication to facilitate tracking.*

NOTE: *Change request status is often denoted as open, under investigation, rejected, postponed, approved for implementation, allocated (i.e., allocated to a developer for implementation), implemented, fixed, closed, etc.*

Assigning a status makes tracking of the implementation of changes a lot easier. Each implementation stage means a new status, by means of which customer and project staff are able to follow the progress of implementation (see status model in SUP.9, figure 2–15).

BP5: Establish the dependencies and relationships to other change requests. *Identify the relationship of a change request to other change requests to establish dependencies, e.g., for all changes to a specific software component or all changes related to a specific software release.*

If several change requests refer to one component or if several change requests are to be implemented in one software release, knowledge of the dependencies is particularly important for reasons of synchronization. A further dependency exists if several change requests (e.g., on different components) are raised as a result of one problem. Without this knowledge it would not be possible to sufficiently prove the successful removal of a problem, for instance, via tests.

BP6: Assess the impact of the change. *Assess the technical impact and potential benefits of the change request.*

NOTE: A Change Request Board (CRB) is a common mechanism used to assess change requests.

In a first step, the technical impacts and the potential benefit of the change request for the project must be identified. This is done in the Change Request Board (CRB, often also called Change Control Board, CCB) meeting. In small projects, the CRB often consists of one person only (e.g., the project manager). In large projects, however, the board is mostly constituted of the project manager, the subproject manager, and other people with comprehensive experience with a good overview and expert domain knowledge of individual subareas (e.g., software, hardware, mechanics, system). As a minimum, the following topics are considered:

- Technical feasibility (e.g., impact on architecture, design, coding, and test activities)
- Potential benefit for the affected subareas

The non-technical factors are identified in BP7.

BP7: Analyze and prioritize change requests. *Change requests are analyzed in terms of resource requirements, scheduling issues, risks and benefits. For every change request a priority is specified that indicates the urgency of the change request to be considered.*

NOTE: For scheduling issues refer also to the product release process (SPL.2).

During the identification of the non-technical factors, possible impacts and risks are considered which may result from the implementation of the change. In addition, the criteria with which the successful implementation is to be confirmed must also be defined. The following, additional issues require detailed consideration:

- Resource requirements/staff effort
- Impacts on the end user
- Impacts on the project (resource situation, process sequence, deadline shifts, work load, etc.)
- Costs or additional costs that may not be provided for in the contract
- Urgency, i.e., the time remaining until scheduled implementation, taking the release plan into account
- Implementation risks

BP8: Approve change requests before implementation. *Change requests are approved on the basis of priority and availability of resources before implementation.*

For each change request, a decision regarding its implementation must be taken and documented, taking into account schedule, staff workload, and consequences identified in the preceding Base Practices. It is not just a simple Yes-or-No decision, but also a decision on the implementation date. In iterative development, individual change requests are normally selected from the number of requests awaiting implementation and assigned to the different iterations/releases (see also BP10).

BP9: Identify and plan the verification and validation activities to be performed for implemented changes. *Before implementing a change the required verification and validation activities to be undertaken are identified and planned.*

Measures must be specified for the verification and validation of changes. Usually, these are tests, trials, or reviews, performed to verify that the changes are correctly implemented according to the prioritization required in BP7. Their purpose is to prove the following:

- The changes comply with the requirements of the customer and the requirements of the market.
- Implementation of the changes was technically defect-free.
- Changes did not cause any undesired side-effects, for instance, that already verified functionality no longer works. To ensure existing functionality, regression testing is performed (see ENG.8 and ENG.10).

BP10: Schedule and allocate the change request. *Approved change requests are scheduled to a particular delivery and are allocated to the resources responsible for implementation, verification and validation*

The changes released in BP8 are planned for a specific target release[92] and passed on to project experts or teams for implementation, verification and validation, using the usual project management methods (i.e., for planning and tracking).

BP11: Review the implemented change. *Changes are reviewed after implementation, verification and validation and before closure to ensure that they have the desired effect and meet their objectives and respective verification criteria.*

A review is performed after implementation, verification and validation of the change to verify that each change has been correctly implemented according to its objective. In general, such reviews are performed by project staff and focus on the following questions:

---

92. A release may include both corrective and adaptive changes.

- Were the measures taken effective?
- Has the problem been permanently resolved?
- Are additional measures necessary to avoid reoccurrence of the problem?

The results of previous tests or other verification measures (e.g., reviews) are taken into account. If the review result is positive, the change is considered successfully completed; if it is not, rework is necessary. Review results must be documented for later evidence.

BP12: Change requests are tracked until closure. Feedback to the initiator is provided.

After a change has been successfully completed (see BP11), the initiator of the change and all involved parties receive explicit feedback according to the defined procedure (BP2), thus ensuring feedback of information.

2.18.4 Designated Work Products

08-28 Change Management Plan

The change management plan defines how changes in the project are recorded and implemented. It describes the lifecycle of a change, starting with the identification of the change, followed by its recording, analysis, description and management, and ending with its implementation. It also defines the possible states a change can pass through. In practice, these things are often defined in a process description at an organizational level, possibly supplemented by project-specific definitions (see also SUP.9 08-27—Problem management plan).

13-16 Change Request

Change requests typically contain the following sections:
- Purpose of the change or change description
- Status
- Initiator, contact information
- Impacted systems
- Impacts on associated documentation
- Criticality, desired implementation date
- Measures for verification/validation

13-21 Change Control Record

The change control record is used to track changes against a particular development baseline. Changed customer requirements and other types of change requests frequently cause a variety of detailed changes on different documents and system components. The change control record specifies the desired change (e.g., via reference to the change request) as well as all the individual changes associated with it. In documents, for example, this can be done in a change history or by means of a CM system. It is important that individual changes can be traced back to the original change request.

A change control record may, for instance, contain the following information (see also SUP.9 13.07—Problem record):

- Reference to the change request
- Systems and documentation impacted by the change, as well as a list of all the individual changes it entails
- Staff responsible for the change(s)
- Change description
- Associated approvals

> **Note to Assessors**
>
> In the assessment, the effectiveness of change mechanisms can best be reconstructed by sample checking of individual changes.

2.18.5 Characteristics of Level 2

Same as SUP.9

2.19 MAN.3 Project Management

2.19.1 Purpose

The purpose of the Project management process is to identify, establish, plan, coordinate, and monitor the activities, tasks, and resources necessary for a project to produce a product and/or service, in the context of the project's requirements and constraints.

Projects are temporary, complex endeavors which are usually subject to challenging deadlines. There is an added difficulty in the automotive industry that due to protracted purchasing negotiations the »official« project start is sometimes too late, for instance, at a time when first prototypes are already available. However, the project completion date (often the start of production (SOP) of a new vehicle), is kept. Moreover, the unclear contractual situation at project start may create problems on both sides: The customer's project manager is unsure if he can now submit requirements, deadlines, etc., to the development organization (supplier). In order to meet deadlines, the development organization must often start preliminary work without a clearly defined requirements specification document, and without a legally binding contract. Usually, the time a project loses at the beginning cannot be regained during its lifetime so that it finds itself on the critical path right from the start. For this reason, systematic project management with sufficiently detailed planning and appropriate mechanisms for project progress tracking is one of the basic preconditions to ensure that critical projects accomplish their objectives at all.

2.19.2 Characteristics Particular to the Automotive Industry

In the automotive industry, development projects usually include the development of hardware, software, and mechanical components. Available hardware, mechanics, and software platforms are often adapted for a particular customer. In larger development organizations, these projects are typically developed by different customer teams/domain departments with different line responsibilities. In complex projects there are additional teams for test and trials, integration, prototyping and platform development (see also our deliberations in the engineering processes).

This is also reflected in the project organization, with projects typically consisting of several subprojects. In most projects, the software is developed by one or several subprojects.

> **Note to Assessors**
>
> Many assessments choose to assess only the software development part of a project. Nevertheless, assessment of the project management process should also consider the interaction of the software subproject(s) with the overall project.

2.19.3 Base Practices

BP1: Define the scope of work. Define the work to be undertaken by the project, and confirm that the goals of the project are feasible with available resources and constraints.

Before establishing a project, the project scope should be defined. According to [PMBOK 2004], the following is needed to determine the project scope:

- The Project rationale/ the business needs
- The project goals, including quantifiable criteria to measure the goal achievement
- A description of the products and/or services to be developed
- An overview of all deliverables to be created

The project scope is often documented in the form of a scope statement[93] and a project work breakdown structure (WBS).

> **Note to Assessors**
>
> In internal development projects (e.g., the development of a development tool), the definition of the project scope is frequently insufficient. The project definition is often missing, and the exact scope of the project is not properly defined (for instance, in case of a development tool: Is the user manual part of the scope? Is initial support part of the project?). Regarding project planning and tracking in internal projects, the same rules and standards are to be applied as in customer projects.

During the course of the project, the BP1 cycle is executed several times. At project start, a rough project work breakdown structure is created. This is further refined at each entry into a new project phase (see BP2), down to the level of work packages and activities (see BP4). In the automotive industry, a project's scope may frequently change during its life time. There are several reasons for this:

- Requirements are not always clearly defined at project start. Due to the way the project progresses and because of modifications to technical concepts (e.g., functional distribution in the vehicle), changes and new requirements arise.
- Due to current market requirements (e.g., a competitor comes out with new functionality, thus gaining a competitive advantage) and current innovations, new requirements and functionality must be integrated that were not known at project start.
- Adherence of production dates (Start of Production, etc.) and associated milestones like product pre-series, sample delivery dates, etc., have highest priority. Functionality is therefore reduced if agreed milestones (e.g., sample versions) are endangered.

A well functioning change management system is therefore of particular importance (see SUP.10).

Project goals must be within agreed cost, time, and quality parameters

- That are specific to the project,
- Measurable or quantifiable,
- Realistic, and
- Available in writing.

Non-goals should also be listed (what is outside the scope of the project?). The conditions needed for the achievement of the goals should also be documented.

---

93. The scope statement is not listed in Automotive SPICE as a separate work product (see also [Hindel et al. 2006]).

2.19 MAN.3 Project Management

In the automotive industry, project goals are usually clearly regulated by contract. For the non-technical scope, goals are, for instance, deadlines for sample versions or number of items to be delivered. The technical scope, e.g., function requirements, is typically described in the technical specifications (see also ENG.1). The main objective is to get the vehicle component with the planned functionality ready for SOP.

In the automotive industry, the product scope is often planned in the form of functions to be implemented by means of a function list[94]. This list is used to plan the content of the individual sample versions and software releases, to allocate functions (possibly also bugfixes) to different software releases and milestones, and to track their implementation. Furthermore, based on this list, the functional scope can be agreed between the customer and supplier at an early stage. The customer knows which functions are planned for which delivery and when they are to be available to him for testing. Figure 2–19 is an excerpt of a function list of a radio navigation system (RNS).

Category	ID	Function Description	Deliveries 4/17 Est.	Deliveries 4/17 Act.	Deliveries 5/30 Est.	Deliveries 5/30 Act.
CD	**CD**					
CD	CD001	Selection internal CD	X			
CD	CD002	CD control function: (play/pause, track search)	X			
CD	...	...				
Audio	**Aud**					
AU	AU001	Source switch FM	X			
AU	AU002	...				
AU	AU003	Sound adjustment: bass	X			
AU	...	...				
MP3	**MP3**					
MP3	MP3001	MP3 control function: (play/pause, track search)			X	

Figure 2–19   Excerpt of an RNS function list

Ideally, the achievability of project goals should be checked prior to project start. In complex projects and where new technologies are involved, a feasibility study is usually conducted. This is often done in the form of a separate (pilot) project. The study of the feasibility typically focuses on technical aspects in relation to cost and benefit considerations (see also [Kerzner 2001]). Often several alternatives are considered, as a result of which the following questions should be answered:

---

94. Often called Feature-by-Phase Matrix or Delivery Plan.

- Which alternative is going to be selected, and is the project technically feasible?
- Which resources are required, which costs are expected to be incurred, and are the required resources actually expected to be available[95]?
- What time frame is needed and are target due-dates at all feasible?
- What are the opportunities and risks?

BP2: Define project life cycle. *Define the life cycle for the project, which is appropriate to the scope, context, magnitude and complexity of the project.*

NOTE: *The consistency between the project life cycle and the car development process should be verified.*

For each project, a suitable and appropriate project lifecycle is specified for the management of the project. At the top level, it typically consists of several project phases. Figure 2–20 shows an example of a lifecycle for an ECU development project.

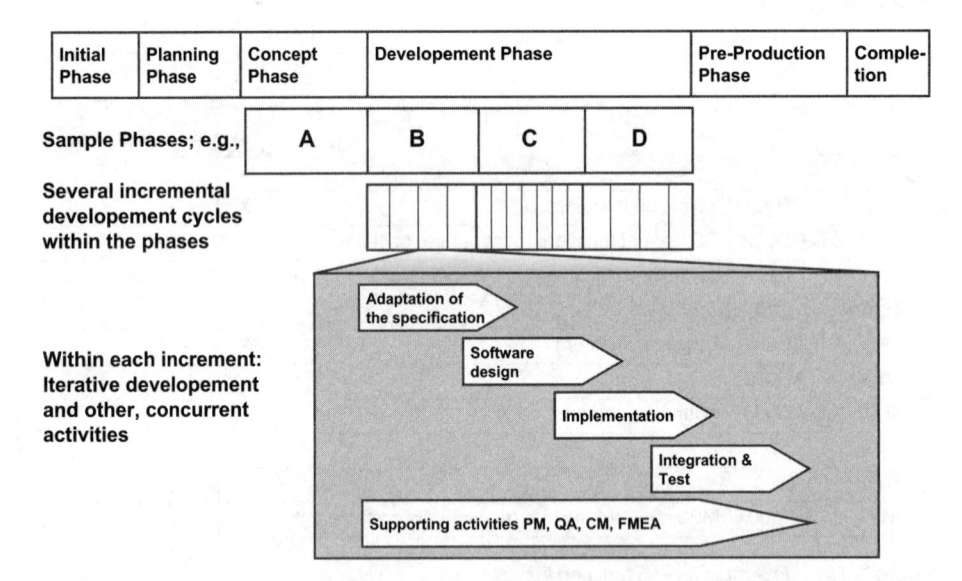

Figure 2–20   Example of a project lifecycle in an ECU development project (supplier view)

Project phases should be separated by milestones. Many organizations have defined phase transitions at defined milestones. There are defined criteria to be used to check if a milestone has been reached and if the phase can be considered completed. Typical milestones during the implementation phase are the sample delivery dates. During the run-up to a delivery this means a review of required

---

95. The same resources are frequently scheduled in several projects. In this case a corresponding risk evaluation is to be carried out and alternatives must be considered.

2.19 MAN.3 Project Management

documentation and verification to see if planned quality checks were completely and successfully performed.

Project phases break a project down into larger segments. In the development of complex products, the actual implementation phase often lasts one to several years. The phases must therefore be refined further. At the lower level, project phases are refined using process models which define the process steps, methods, and work products. In the automotive industry, development cycles usually run through several iterations of the V-model (for more details, see [V-model XT]).

Up to a certain degree of detail, processes and milestones are often dictated by the customer's own product development methodology. For example, a vehicle manufacturer's development processes, including so-called sample phases and resulting development prototypes, can set a very tight milestone time-frame that profoundly impacts the development processes of his development partners.

In the course of this base practice, it needs to be defined which of the process model's project phases, milestones, development cycles, and process steps are to be performed. This includes specifying whether 2 or 3 development cycles are to be performed during the B-sample development and whether each increment will cycle the »small V« (according to the V-model) completely.

Moreover, the extent to which the supplier's project development lifecycle and the OEM's vehicle development process interconnect and fit each other should also be examined. Figure 2–21 illustrates typical milestones and associated development phases.

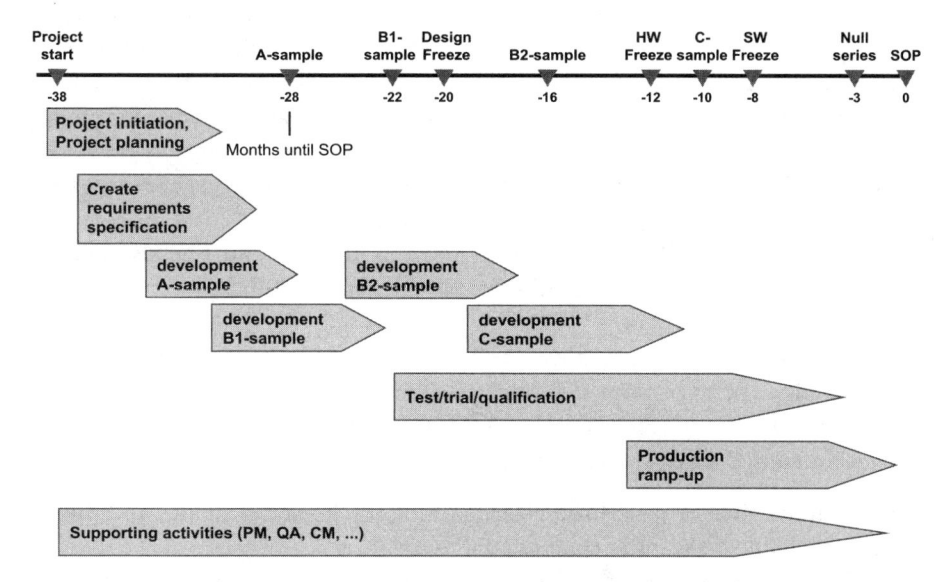

Figure 2–21   Milestone grid and associated development phases

BP3: Determine and maintain estimates for project attributes. *Define and maintain baselines for project attributes.*

NOTE: *Project attributes may include 1) business and quality goals for the project, 2) resources for the project and 3) project effort, schedule and budget.*

NOTE: *Appropriate estimation methods should be used.*

NOTE: *A development strategy is determined and resources for the development life cycle to satisfy requirements are estimated.*

NOTE: *Resources may include required infrastructure and communication mechanisms.*

NOTE: *Project risks and quality criteria may be considered when estimating project attributes.*

The primary project parameters (effort, costs, deadlines and quality) are estimated (see also [Hindel et al. 2006]). These may include the following:

1. Effort Estimation: The effort required for the creation or implementation of the work package or module (e.g., in person days) is estimated, based on the project's work breakdown structure. Effort is frequently estimated bottom-up (»micro estimation«), breaking the work to be done down into small work packages. At the lowest level, it is estimated on a per-work package basis and then added up step-by-step to calculate the total estimated project effort.
2. Schedule Estimation: Based on the estimated effort and available resources, target dates are estimated (see BP6). In the automotive industry, sample delivery dates and SOP are typically fixed, so planning focus is on intermediate deadlines and the functional scope expected to be ready at individual milestones.
3. Cost Estimation: Resource costs are determined by multiplying the estimated work effort with associated staffing cost rates. Additional costs for required resources, like development environment and material, must be also added. This point is not to be confused with the calculation of the quotation for the project.
4. Quality Planning: The quality of intermediate and end products must be planned (e.g., maximum defect rates allowed for particular development versions, degree of test coverage). This is usually done during, or in coordination with, the quality planning (see SUP.1 BP1 and BP3).

Organizations often only perform effort, cost, and schedule estimates and find it difficult to define measurable quality goals.

One of the notes above refers to the use of estimation methods. In practice, expert estimation techniques like Delphi and 3-point estimates are frequently used, whereas advanced methods like Function Point or COCOMO are hardly used at all in the automotive industry. Informal expert estimates are fairly common practice in projects. Typically, results are undocumented estimates whose derivation is unclear. Estimation templates have proved to be practicable for use in projects as they allow structured estimates with relatively little effort. For further deliberations regarding estimation methods, refer to [Hindel et al. 2006].

It is of advantage (though not required) to record actuals in such a way as to allow the creation of, for instance, standard estimation templates with typical, pre-defined work package efforts based on collected experiences. These experience values are then used in new projects (so-called historical data). In most cases the accuracy of the estimate is thus considerably improved. It presupposes that standard work packages can be defined that are applicable in all projects.

Another possibility to gather experience and improve estimates is to conduct project completion reviews or »Lessons Learned«. There, valuable experiences are gathered which can be used in future projects. Especially in projects with long duration it makes sense to perform several completion reviews during the course of the project (for instance, at the end of important project phases). In this case, the objective is no longer just to learn for future projects but to directly apply the lessons learned in the next phase of the same project. Identified improvements, for example, relate to adaptations of processes and project descriptions, templates and checklists, as well as measures to improve team work. Regarding estimates, the following experiences can be gathered:

- Preparation of the actual work package efforts to be used for future projects (for instance, in the form of a historical data base)
- Improvements of estimation models based on actual efforts
- Follow-up of project risks: Consideration in future estimates

Estimates are made at different times during the course of the project. Rough estimates are made at the beginning of the project or, for instance, included in the quotation process. During the project planning phase, these rough estimates are adjusted and particularized. In the end, estimates are made at the level of work packages. Further detailing and adjustments are made during the implementation phase, for instance, during entry into a new project or sample phase, or when replanning is performed. Automotive SPICE requires that these estimates (and the associated estimate documentation) are compiled in the form of baselines. These baselines can be part of an overall project baseline (see SUP.8).

Maintaining estimates means that changes during the course of the project are taken into account, with the consequence that estimates are repeated or that estimate documentation is adjusted whenever needed. Estimates can also consider risks and may, for instance, take risk allowances into consideration.

> **Note to Assessors**
>
> Informal expert estimations are widespread. There, the estimate is done rather informally by staff with expertise in the corresponding domain area. In this case, the results of informal estimates are to be traceably documented. Whenever possible, the estimate should be done by at least two experts or, if only one conducts the estimate, be reviewed by a second expert.

BP4: Define project activities. *Plan project activities according to defined project lifecycle and estimations, define and monitor dependencies between activities.*

NOTE: *The activities and related work packages should be of manageable size to ensure that adequate progress monitoring is possible.*

The project work breakdown structure (WBS, see BP1) is defined in more detail. Activities are derived from the lowest level of the WBS. Furthermore, activities are added which are derived from the project lifecycle (see BP2). Reference is given in a note that the planned work packages of the project need to be of manageable size to ensure that progress monitoring is possible. This depends on the duration, size, complexity, reporting cycle, etc., of the project. In projects with a completion time of more than one year with monthly project tracking, one reference value may, for example, be a maximum of 4 weeks or 160 person-hours of effort. A rule of thumb says that work packages should be detailed to such a level that their progress can be monitored within the scope of one reporting cycle.

Based on the lifecycle model (e.g., iterative model), different iterative instances of activities are created, for example, design of 1st prototype, design of 2nd prototype, and so on.

Dependencies between the activities are defined. For instance, a design of a particular component must be created before implementation can take place. The concrete chronological order is identified in BP6.

BP5: Define skill needs. *Identify required skills needed for the project and allocate them to individuals and teams.*

All necessary qualifications are identified on the basis of the tasks to be completed and compared with the qualifications of candidate staff members. Based on this, the optimal staffing structure of the project team is planned. This is typically done by the project or line manager. If designated staff do not have the necessary qualifications, suitable measures must be taken to train them.

2.19 MAN.3 Project Management

BP6: Define and maintain project schedule. Allocate resources to activities and determine schedule for each activity and for the whole project.

NOTE: *This includes appropriate re-planning.*

NOTE: *Project time schedule has to keep updated during lifetime of the project continuously*

The activities planned for the project are assigned to individual, qualified members of the team. It must be known which of the differently qualified staff are available from when to when and to what extent, and what the dependencies are between the activities. The effort needed for each activity is also known from the effort estimate. This information is used to allocate resources, identify the actual duration of activities, and to determine the start and end dates. As a result, a documented schedule is created.

Maintaining schedules means that changes during the course of the project are taken into account, and also, that the schedule is put under configuration control and republished and redistributed after each change. The schedule is adjusted whenever changes have an impact on subsequent milestones or deadlines (e.g., if deadlines of important activities or milestones need to be shifted). The project should determine the type of changes that are relevant for schedule adjustments.

In addition, the schedule must be regularly reviewed and updated. The updating frequency depends on the actual project situation and the schedule's degree of detail. Typical are weekly updates; as a minimum, there should be one update per month.

> **Note to Assessors**
>
> Schedules are progressively refined during the course of the project. At the beginning of a long-term project, detailed planning of all tasks is impossible. In iterative development, however, at least the current cycle should be planned in detail. For later cycles, rough planning is sufficient. However, details must be planned no later than at the beginning of a new cycle, (see also BP4 for references regarding manageable size of activities) and the rough plan of the subsequent cycle must be updated.
>
> If deadlines need to be shifted, care must be taken to consider dependencies in the plan, update subsequent dates, and identify associated risks.

Since development projects are usually subject to enormous time pressure (the problem is time-to-market), schedules always carry corresponding risks. These risks can be identified within the context of MAN.5 (risk management), be minimized by measures, and be regularly reassessed and tracked.

BP7: Identify and monitor project interfaces. *Identify and agree interfaces of the project with other (sub) projects, organizational units and other stakeholders and monitor agreed commitments.*

NOTE: *The project planning and monitoring may include all involved parties like quality assurance, production, car integration, testing and prototype manufacturing.*

Agreement must be reached with stakeholders regarding the handling of interfaces. Stakeholders include the responsible line management, the customer, internal teams such as QA and production, test, and integration departments, prototyping, the purchasing department, sales, marketing, product management, and so on. The following questions need to be settled: What information is exchanged, how and when is it exchanged, how is the work status to be reported and tracked, are there joint meetings, how are these people and groups included in the communication flow, and so on. Adherence to these agreements must be actively monitored within the project management process.

BP8: Establish project plan. *Collect and maintain project master plan and other relevant plans to document the project scope and goals, resources, infrastructure, interfaces and communication mechanisms.*

The project plan consists of one or several planning documents that define the project scope and relevant features of the project, such as project goals, scope of work, activities and tasks, resources and interfaces, and so on. Automotive SPICE distinguishes between the overall project plan and additional plans. Additional plans can be plans that are created in subprojects or within the context of other processes. Examples are the quality plan and the configuration management plan. In this case, the overall project plan combines the individual sub plans and uses common milestones to synchronize them. In smaller projects, there is typically just one single project plan.

> **Note to Assessors**
> If a project consists of several subprojects, these subprojects often maintain separate plans. This is perfectly acceptable. However, it needs to be ascertained during the assessment that the individual subproject plans are compatible with the overall project plan (e. g., via common milestones), and that they are regularly adjusted.

The overall project plan must be updated and adjusted as the project progresses. Rules should be defined regarding how the plans are to be updated; for instance, at regular intervals or after major changes. In the latter case, specification is needed as to what constitutes a major change.

One aspect which is recommended but not required by Automotive SPICE at Level 1 is that planning baselines are reviewed and subsequently released by management.

BP9: Implement the project plan. Implement planning activities of the project.

Implementation of the project plan includes making the plan available to all project team members (for instance, distributed via e-mail), and also that they are familiar with it and understand what they are supposed to do when. This may be done in the form of a kickoff event or some other kind of meeting. If project team members do not start activities independently and in time, they may need to be triggered by the project manager, for instance, in project meetings. Otherwise, activities are carried out according to plan. Refer to BP11 regarding the topic of progress monitoring.

BP10: Monitor project attributes. Monitor project scope, budget, cost, resources and other necessary attributes and document significant deviations of them against the project plan.

The progress of the project parameters (quality, actual effort, costs, schedule adherence, etc.) is identified and documented. Planning deviations are recorded. Refer to BP11 regarding the topic of progress monitoring; for the topic of frequency refer to BP7.

In most development projects, a lot of technical questions and problems are to be clarified and resolved during the course of the project. Resolution of these questions must be systematically controlled and tracked, for instance, by means of an open issues list (OIL, see figure 2–3). Progress monitoring may also include tracking and renewed evaluation of risks (see MAN.5 BP6).

BP11: Review and report progress of the project. Regularly report and review the status of the project against the project plans to all affected parties. This includes reports to the car producer. Regularly evaluate the performance of the project.

NOTE: *Project reviews may be executed at regular intervals by the management.*

The actual work status is compared with the plan. Recognized deviations from the current planning baseline are documented. This can be done in project status reports. Progress is regularly reported (see work product 15-06—Project status report). In practice, progress monitoring usually takes place at two levels:

- Progress monitoring at activity level: Progress of activities is monitored at short intervals (e.g., weekly). At the lowest level, developers report the progress to the next level up. In smaller projects this is the project manager, in larger projects there are some in-between levels (e.g., team manager or subproject manager). Used methods are, for instance, team meetings, the identification of actual effort and degree of completion, or of actual effort and residual effort, respectively. This is the raw data needed for progress monitoring at project level.

- Progress monitoring at project level: Here, progress data becomes successively aggregated from the lower levels of the hierarchy to the higher ones to help the project manager gain a clear picture of the progress at the overall project level. This he can then communicate to external parties (management, customer). The results are (e.g., monthly) documented in project status reports. Applied methods are project reviews, metrics for project progress tracking, milestone trend analyses, and the earned-value method (see [Hindel et al. 2006]). A note points out that regular progress monitoring at project level can also be done by (line) management. In critical, complex, or large projects, there may exist a common steering committee comprised of management representatives of both customer and supplier.

Besides progress reports, project meetings form the most important means of project control. Figure 2–22 lists some frequently occurring types of project meetings.

Type of meeting	Purpose	Participants
Team meetings	Identify problems and solutions at activity level, work status	Project manager/team leader, developer
Internal progress reviews	Work status compared to plan, measures, designated problems, risks, change requests	Project manager/team leader, developer
Formal progress reviews	Work status compared to plan, in brief: selected problems, measures, risks	Project manager/team leader, management
Milestone reviews	Work status compared to plan, check of formal requirements, release of the next phase	Project manager/team leader, management, QA, possibly customer representative
Steering committee	Report and accounts, designated problems, strategic questions, coordination of different interests, taking of important project decisions	Project manager/team leader, management, customer representative

Figure 2–22   Types of project meetings

BP12: Act to correct deviations. *Take action when project goals are not achieved, correct deviations from plan and prevent recurrence of problems identified in the project. Update project plans accordingly.*

If during the execution of the project and during the monitoring of its progress, deviations against the plan are identified to the effect that project goals cannot be reached, appropriate measures must be initiated. Several different alternatives are possible:

2.19 MAN.3 Project Management

- Perform sequential activities along the critical path in parallel
- Change staff/more staff/more experienced staff
- Outsource work to give project members time for other tasks
- Change or downsize scope of work/functionality
- Schedule delay, open dialog with the customer

These alternatives may carry considerable risks and additional costs and may, under certain circumstances, have exactly the opposite effect. Another, unfortunately frequently practiced alternative is to cut corners and compromise quality (e.g., fewer reviews, no tests or shorter tests). It is strongly advised not to do this.

Generally, corrective measures lead to changes of the plan; this is why the project plan must be adapted accordingly. Larger changes of the plan may make it necessary to conduct another planning review.

In addition to the implementation of measures for the correction of deviations from the plan, Automotive SPICE also requires measures for the removal or avoidance of causes.

> **Note to Assessors**
>
> It is simply impossible to devise immediate measures for every problem to prevent its recurrence; sometimes it is simply too late to be able to remove the causes in the project, because the conditions that led to the problem can no longer be eliminated (for instance, if there are resource problems and if short-term outsourcing is impossible, or if new recruitment cannot be made). It is important in these cases to conduct a root cause analysis and to adequately consider the situation (e.g., at project meetings, or as a part of the »Lessons Learned« evaluations).

2.19.4 Designated Work Products

08-12 Project Plan

The project plan consists of one or several planning documents that define the project scope and the primary features of the project. The project plan is the basis for project control. If the project plan consists of several planning documents, care must be taken to ensure that, in sum, the individual documents form a coherent and cohesive whole. However, the project plan does not necessarily need to be a coherent document. It is often implemented in the form of a directory structure with different files. The following table provides an exemplary structure of a project plan.

Exemplary Structure of a Project Plan for a Large Project

1. Brief Description of the Project Objectives and Project Scope
    1.1 Brief Description of the Project
    1.2 Background
    1.3 Objectives
    1.4 Project Scope
    1.5 Project Budget
    1.6 Project Runtime
    1.7 Prerequisites and Commitments
    1.8 Basic Planning Assumptions
    1.9 Basic Decisions
    1.10 Interfaces
2. Requirements Specification and System Specification
3. Project Work Breakdown Structure and Project Schedule
    3.1 Overview of the Work Packages
    3.2 Project Work Breakdown Structure (Overview)
    3.3 Schedule (Overview)
    3.4 Description of the Work Packages
         WP 1 ...
         WP 2 ...
         WP *n* Project Management
    3.5 Summary of Effort Estimations
4. Roles and Responsibilities
5. Risks
6. Used Processes and Tools
7. Reporting
8. Basic Rules for the Project Team
9. Project Team
10. Project Provisions
11. Additional Information
    11.1 Change process
    11.2 Documentation
    11.3 Joint Documents
12. Definitions, Terms, Abbreviations
13. Appendix: Effort Estimation of Required Staff
14. Appendix: Human Resource Planning
15. Appendix: More Detailed Project Work Breakdown Structure
16. Appendix: More Detailed Schedule

13-16 Change Request

See SUP.10

14-06 Schedule

The schedule should contain (see also BP6):

- All activities to be performed, sequence, and dependencies between them
- Milestones
- Duration and estimated effort of the activities, start and completion dates
- The assignment of resources to activities

Furthermore, the critical path or paths should be identified in the schedule. Results of schedule tracking are typically documented in the plan (for instance, the degree of completion, activities completed). Often, the schedule is maintained using tools like Microsoft Project, R-Plan, and so on. For simple schedules, however, Microsoft Excel may be sufficient.

14-09 Work Breakdown Structure

The project work breakdown structure (WBS) arranges and defines the project's overall content and scope. WBS is the tool that breaks down the project scope into »manageable« items. Each lower level contains a more detailed description of the project items of the level above. Work packages constitute the lowest level of the WBS. Deliverables that are not included in the WBS are outside the project scope. At the top level, the WBS should be structured according to project phases or main deliverables. Some alternative forms of description are possible: Often a structure diagram is chosen, but list descriptions with numbering and indentations, and mindmaps, are also common.

In addition to the product and product components to be developed, other deliverables of the project comprise the user manual, test documentation, project status reports, trainings, etc., as well as deliverables derived from the project lifecycle[96] (e.g., documents which are not passed on to the customer; for instance, design documents).

> **Note to Assessors**
>
> Particularly smaller projects often do not create a separate document for the project work breakdown structure. The WBS is then implicitly included in the project schedule.

---

96. Project phases consist of chronologically connected project activities that are logically interconnected. A project typically passes through the following phases, which again may be made up of several subphases: start phase, planning phase, implementation phase, and a final phase.

15-06 Project Status Report

The project status report should include the following information:

- Cost situation (current development costs or production cost per item, anticipated cost development, etc.)
- Schedule situation (anticipated finishing dates, trends; for instance, regarding achievement of the next milestones, etc.)
- Scope of work/functionality
- Quality situation (defect situation, customer relevant problems, etc.)
- Resource status (e.g., human resource bottlenecks)
- Risks (summary of risks)
- Designated (often technical) problems
- Initiated measures. Who needs to act?

In organizations which need to monitor a larger number of projects, project status reports are highly aggregated and presented in a standardized form, for instance, as performance indicators based on »traffic lights«. (see figure 2–23).

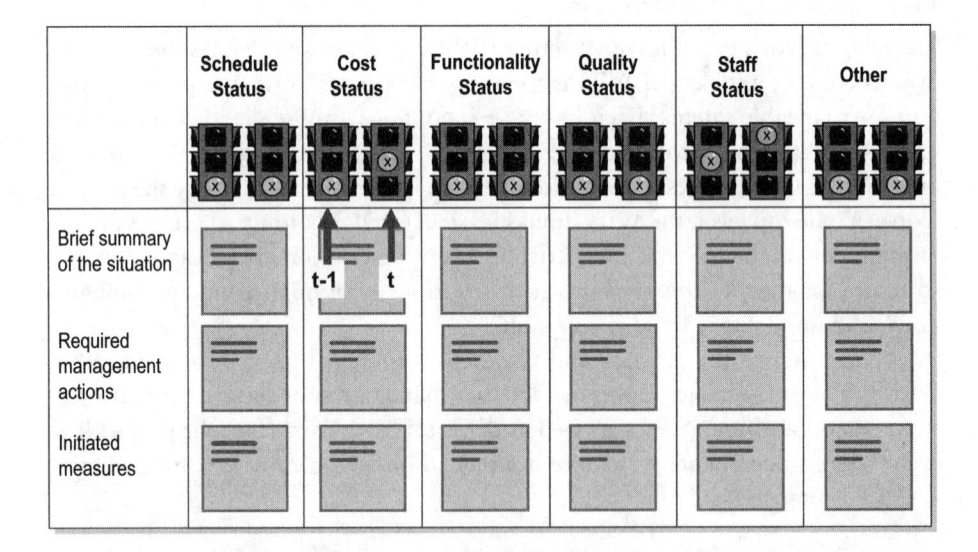

Figure 2–23   Example of a highly aggregated project status report for higher management levels

2.19.5 Characteristics of Level 2

Project Management of the Project Management Process

At Level 2 (see PA 2.1, control of the process performance), basic project management principles are also required, but now they are applied to the respective process (and not the project).

The process is required to be planned based on targets, and adherence to the plan is monitored. For MAN.3 this means that project management activities like the creation of a project plan must be planned and tracked. This applies also to activities, such as project meetings, detailed project planning and planning updates, progress monitoring, the creation of progress reports, and so on.

For instance, it makes sense to plan the initial completion date of the project plan and the updates that need to be made at the start of new phases. In addition, a resource plan for project management activities must also be drawn up, taking recurring activities into account. This is particularly important if the project manager has to perform technical tasks on top of his managerial ones. This typically happens in smaller projects. The result is that in stress situations, project management work is frequently neglected.

On Work Product Management

Planning documents change relatively frequently, provided that the documents »live« (i.e., they are actively used as planning and control instruments). This is why planning documents cannot be reviewed after each minor change. However, the project plan and schedule should at least be reviewed whenever planning baselines are done.

2.20 MAN.5 Risk Management

2.20.1 Purpose

The purpose of the Risk management process is to identify, analyze, treat and monitor the risks continuously.

In this context, a risk is understood to denote an unwanted effect or potential problem which may, with a certain probability, occur sometime in the future. Risk occurence is accompanied by damage to the project, i.e., it has a negative effect on project objectives and may, for instance, cause cost increases, deadline shifts, quality problems, or other types of damage. The objective of risk management is to identify risks and to avoid or alleviate the probability of occurrence and/or the impacts of risk occurrence as much as possible. In this context, the term »risk« is closely related to the term »problem«. The decisive difference is that a problem has either already occurred or that it will occur with absolute certainty.

Risk management methods allow the identification, analysis, and evaluation of any kind[97] of risk, and enable risk monitoring throughout the entire project lifecycle.

---

97. Including non-technical risks, in contrast to the FMEA.

2.20.2 Characteristics Particular to the Automotive Industry

The FMEA[98] method (Failure Modes and Effects Analysis) constitutes a special form of risk management. In automotive development, the implementation of system FMEAs is mandatory.

An FMEA identifies and assesses anticipated failure sources in the design (so-called construction FMEA[99]) or in production (so-called process FMEA). The methodology of the system FMEA combines these two perspectives. FMEAs thus concentrate on risks related to the product and manufacturing process and are primarily used in early project phases after availability of an initial design (or concept of a production process). ECU software must of course also be considered in the FMEA.

Risk management, as described here, additionally addresses all possible further risks (such as project risks, development process risks, and organizational risks). If components are developed by suppliers, risk management can take place at several levels in the project:

- At the level of the development project (both on the side of the supplier and the customer)
- At the level of the vehicle project on the side of the customer
- At an organizational level (customer as well as supplier organization)

> **Note to Assessors**
>
> The scope of application of the risk management process to be addressed must therefore be specified in the context of the process assessment.

2.20.3 Base Practices

BP1: Establish risk management scope. Determine the scope of risk management to be performed for the project, in accordance with organizational risk management policies.

On this practice, refer to the deliberations at the end of the previous section. Otherwise, the scope typically includes:

- Product components
  This includes the definition of the product components for which the development teams are to apply risk management measures.

---

98. FMEA can be combined with other analysis techniques, for instance, the FTA (fault tree analysis). FTAs analyze the causes of failure modes) and ETA (event tree analysis. ETAs analyze the possible consequences of failure modes).
99. If the principle of the construction FMEA is applied to software, we talk about a software-FMEA. Software parts are often addressed as part of an overall FMEA.

- Project phases
  Normally all project phases are considered in which risks may occur. However, this may be different if phases are handled by different teams.
- Product lifecycle phases
  Besides the development phases, the product lifecycle also comprises phases like start-up, operation, putting out of operation, and so on. Certain phases may possibly be excluded and delegated to other (relevant) teams that need not necessarily be part of the project.

The risk management scope must be in compliance with policies at the organizational level. For instance, the following may have been agreed at the organizational level:

- The project types for which risk management must be performed, i.e., all projects or only projects of categories A and B.
- Allowed tailoring within risk management (see Level 3); e.g., projects with a project volume larger than n USD must apply the complete risk management process. Smaller projects need only apply a part thereof.
- Exclusion of certain risks. For example, production risks could be excluded in a development project, since they are covered by the risk management applied by the production department.

BP2: Define risk management strategies. Define appropriate strategies to identify risks, mitigate risks and set acceptability levels for each risk or set of risks, both at the project and organizational level.

Risk management focuses on the project. One of the project tasks is to identify, analyze, and mitigate risks. Risk tracking, however, is done at project and organizational levels. This means that the project team not only tracks the status of risks and associated counter measures, but also regularly reports to the organization (i.e., to management). This is why the base practice requires definition »at both the project and organizational levels«. However, although it makes sense to establish consistent methods and risk indices at an organizational level, it is not required here (only from level 3 onwards). A risk management strategy can cover the following aspects:

- The methods and tools that can be used for the management of risks (risk workshops, risk lists, decision trees, etc.)
- The way in which frequently occurring sources of risks are recorded (e.g., in the form of checklists and catalogs)
- The way risks are organized, categorized, and consolidated
- The way risks and measures are tracked, e.g., using risk indices, etc.
- The time intervals when risk tracking is performed

In less complex projects[100], the strategy could be summarized as follows:

- Risks are identified and collected in the team, whereby risks stated by team members are recorded in a list. Expedient aids are the product structure, the project plan, or risk checklists.
- Risks are analyzed using risk indices (see BP4).
- Risks are managed by planning counter measures (see BP5).
- Risks and associated counter measures are tracked according to specified rules, and regularly reported.

Moreover, Automotive SPICE requires the definition of acceptance thresholds. According to Process Outcome 6 (refer to the original Automotive SPICE text), »risk thresholds« can refer to priorities, probabilities, and consequences. A definition is therefore needed, stating which risks are accepted and from which threshold onwards measures are to be taken (e.g., risks from a risk priority number 4, according to figure 2–24).

BP3: Identify risks. Identify risks to the project both initially within the project strategy and as they develop during the conduct of the project, continuously looking for risk factors at any occurrence of technical or managerial decisions.

NOTE: *Examples of risk areas, which should be analyzed for potential risk reasons or risks factors, include: cost, schedule, effort, resource, and technical.*

NOTE: *Examples of risk factors include: unsolved and solved tradeoffs, decisions of not implementing a project feature, design changes, lack of expected resource.*

It is advisable to identify risks in several steps, for example:

- Prior to the actual project start as part of the decision making process to see whether the project is feasible or not, and whether it should be offered to a customer (also refer to the feasibility analysis in MAN.3 BP1). Early risk identification and analysis may, for instance, result in the decision to forego the project due to high risks, or to reflect associated extra costs in the project calculation.
- If the project has already started, and if the project management and the project team are already committed, a more detailed investigation and analysis is conducted, resulting in concrete counter measures.
- Further risks may be detected during the course of the project which could not be anticipated at the beginning. Each change in the course of the project may bear risks and must be analyzed with regard to them. It is also advisable

---

100. In larger projects with hundreds of members of staff, many different teams identify risks separately. More complex methods are necessary in these cases, for instance, to combine and concentrate risks and status from the team level to higher levels and up to project management, or to determine a risk status for the entire project which can be reported to management.

2.20 MAN.5 Risk Management

to renew risk identification from time to time (e.g., when entering into a new project phase), and to extend and regularly update current risk estimates. This becomes more and more important the longer the project runs.

Moreover, Automotive SPICE requires renewed risk identification after each technical[101] or management decision, for instance, when new deadlines have been agreed.

Risk identification uses different types of input. Examples are schedule, resource planning, cost plan, and technical concept. Besides those factors stated in the notes, frequent risk factors in automotive development projects are unclear requirements, changes of the market situation[102], new and unknown suppliers, delays of important decisions, late project start, as well as unclear responsibility assignment, especially in the case of distributed development[103].

BP4: Analyze risks. Analyze risks to determine the priority in which to apply resources to mitigate these risks.

NOTE: *Issues to be considered in risk analysis include the probability and impact of each identified risk.*

Possible resources are manpower requirements or monetary expenditures, or the provision of infrastructure (e.g., better tools, more test benches). This concerns resources for risk tracking and counter measures. The reason why prioritization is particularly necessary is because even in small projects often dozens of risks are identified. It is simply impossible to track all of them with the same intensity.

The risks are analyzed using risk indices. Typical indices are probability of occurrence and impact. These two indices are multiplied with each other and result in a risk priority number. This number allows the mapping of risks on a one-dimensional scale so that they can be prioritized. A popular form of description is the risk portfolio (see the example in figure 2–24), in which probability levels and impact are defined (in the example each from 1 to 5, whereby 5 is the highest rating). Depending on these, risk priorities are defined, also each from 1 to 5 in the example (this is shown by grayscales in figure 2–24), whereby 1 has highest priority.

---

101. This refers to far-reaching decisions like design changes or decisions regarding non-implementation of functional features, and associated risks.
102. For instance, new functions or competing products, new insights from market studies, findings from competitor analyses, etc.
103. For example, in case of distributed development across several (international) sites, or in case an entire system is developed by several suppliers.

	5	4	3	2	1	1
Probability of occurrence	4	4	3	3	2	1
	3	4	4	3	3	2
	2	5	4	4	3	3
	1	5	5	4	4	3
		1	2	3	4	5
		Impact					

Figure 2–24 *Example of a risk portfolio description*

For an efficient analysis it is advisable to use evaluation levels in connection with risk indices (e.g., small/medium/large, depending on the impact), and to explain them in a table. A frequently encountered difficulty is the establishment of a common understanding among team members of the evaluation levels used for the impact. Since »impact« arises in different categories (time, quality, development costs, cost per item, etc.), the assessment levels must be defined for each of these categories (»From which level of increase in cost per item onwards do we talk about high impact?«). Figure 2–25 illustrates different impact categories for impacts on schedules.

	Impact on Scheduling
	Occurrence of risk will probably result in a ...
1	Schedule delay **< 2 weeks**, Important milestones are not shifted
2	Schedule delay **up to one month**, Important milestones may need to be shifted
3	Schedule delay **1 to 2 months**, Important milestones are definitely shifted
4	Schedule delay **3 to 4 months**, Important milestones are definitely shifted, SOP jeopardized
5	Schedule delay **more than 4 months**, SOP highly jeopardized

Figure 2–25 *Examples of impact categories for impacts on schedules*

BP5: Define risk treatment actions. For each risk (or set of risks) define, perform and track the selected actions to keep/reduce the risks to an acceptable level.

Counter measures are defined and initiated[104]. Common types of counter measures are:

- Avoidance: Risks are avoided by applying changed practices in the project (»preventive measures«).
- Transfer: Risks are transferred to a third party (e.g., subcontractor). Often, transfer addresses only part of a risk. A cost risk can, for instance, be passed on, but associated risks regarding schedule delay or customer dissatisfaction remain.
- Mitigation: Early measures reduce probability of occurrence and/or impact.
- Acceptance: The risk is accepted, e.g., because counter measures are not possible or uneconomical. In this case, it may be advisable to prepare a »contingency plan« for the event of occurrence, or to plan reserves (in the form of provisions, time, resources, etc.).
- Contingency measures: Measures that may be applied in case of risk occurrence are conceived in advance.

Due to the often rather high number of risks, it is usually impossible to apply active counter measures for all the risks. Many small risks must simply be accepted, i.e., the counter measures consist in the acceptance of these risks in conjunction with reserves of time, finances, or personnel.

BP6: Monitor risks. For each risk (or set of risks) define measures (e.g., metrics) to determine changes in the status of a risk and to evaluate the progress of the mitigation activities. Apply and assess these risk measures.

Risk tracking means that risks are regularly discussed during (selected) team meetings and that their assessment and status are updated (for instance, in the form of the risk indices according to BP4). Progress and effect of the counter measures must also be tracked with the help of indices. This can be done by measuring the degree (in percent) of progress as part of project tracking, provided the counter measure is listed as an activity in the project plan. Alternatively, status tracking can be done in the form of traffic lights (Green/Yellow/Red) to indicate whether the counter measure takes an effect as planned or if there are difficulties.

At the overall project level, normally only a few (e.g., 10) high-priority risks are intensively tracked as individual issues. Other risks are delegated to subprojects or they are considered less often, with less effort, and/or in sum.

---

104. In correspondence with the requirements of Level 2, a measure is treated like any other activity in the project, i.e., responsibility is assigned, and it is planned, tracked, etc.

BP7: Take corrective action. When expected progress in risk mitigation is not achieved, take appropriate corrective action to reduce or avoid the impact of risk.

NOTE: Corrective actions may involve developing and implementing new mitigation strategies or adjusting the existing strategies

Counter measures defined in the analysis must be verifiably implemented. If there are indications that they cannot achieve the expected success, measures already taken must be intensified, or new measures must be initiated. If measures generally turn out to be inappropriate or if a risk is identified too late, it is still possible to limit or minimize the damage with the help of a contingency plan containing measures to be applied in the event of occurence.

> **Note to Assessors**
>
> Concerning this process, at least the following questions should be asked:
> - Were risks identified at project start?
> - Were risks completely and comprehensibly documented?
> - Were counter measures planned in an appropriate way, and is this reflected in the schedule and resource plan?
> - Are risks, counter measures, and corresponding planning regularly tracked?
> - Were (in case of longer project run times) further risks identified during the course of the project?

2.20.4 Designated Work Products

Risk List[105]

The risk list is the basic tool of risk management. In it all risks are listed, assessed and prioritized. It often also includes required counter measures, associated dates, responsibilities, status, etc.

---

105. In Automotive SPICE, the risk list is not listed as a separate work product but as part of the »risk management plan« (WP-ID 08-19).

2.20 MAN.5 Risk Management

> **Exemplary Structure of a Risk List**
>
> 1   Risk
>
> - Risk ID
> - Risk Specification
> - Risk Description
> - Probability of Occurrence
> - Extent of Damage
> - Risk Priority Number
>
> 2   Counter Measure
>
> - ID of Counter Measure[a]

a. The measure, for instance, is tracked via the project's open issues list, and a reference to its ID is given.

Besides the risk list, the following basic items are recommended for effective and efficient risk management, even though they are not specified as explicit work products in Automotive SPICE:

- Risk checklists may serve as aids for efficient risk identification and may, for instance, be initially extracted from risks of real projects to be continuously refined and expanded later.
- Risks checklists based on experiences from earlier projects. Their results are consolidated and made available, using »Lessons Learned«
- The set of risk indices to be used by all projects, with comprehensive comments regarding the evaluation levels of the risk indices (see BP4)

2.20.5 Characteristics of Level 2

On Performance Management

Whereas larger projects usually create their own separate risk management plan, planning in smaller projects is often done in different documents:

- Individual actions like risk workshops (during which risks are identified and analyzed, and during which counter measures are planned) occur in the project plan.
- Individual counter measures are planned and tracked in a risk list or in an »open issues list« acting as a central planning and tracking instrument of the project.

On Work Product Management

The requirements of Process Attribute PA 2.2 particularly apply to the risk list and associated measures. If a separate risk management plan is established in larger projects, it needs to be reviewed.

2.21 MAN.6 Measurement

2.21.1 Purpose

The purpose of the Measurement process is to collect and analyze data relating to the products developed and processes implemented within the organization and its projects, to support effective management of the processes and to objectively demonstrate the quality of the products.

»You can't control what you can't measure!« (Tom DeMarco)

Measurement is a key item for successful management in every engineering discipline, and an effective tool for project control. In this context, measurement is understood as a continuous process in which metrics are defined for the processes and products to be measured, and where measurement data are collected, analyzed, and evaluated. On the one hand, the objective of the measurement process is to understand, monitor, control, and optimize processes in order to reduce development effort and costs. Yet on the other, it is to improve the work products under development. Measurement is an important means for project management, and it aids risk management[106] and the assessment of project progress by means of progress measurements (Is the project halfway to completion, or only 30 percent?).

It is important that measurements are done in a goal-oriented way. A measurement program must reflect and support the business goals of an organization. A common approach towards goal-oriented measurement is the Goal/Question/Metric method (see the subsequent excursus on the »GQM method«).

Establishing a measurement process in an organization requires not only the participation of involved staff and management but also associated central resources (for example a measurement team) that institutionalize the process and support involved persons during measurement.

Improvement projects are frequently advised to wait with the introduction of MAN.6 and to first focus on processes such as project management, quality assurance, requirements management, configuration management, and the engineering processes. In our experience this is not advisable. Continuous measurement should already be started at the beginning of an improvement project to be able to show accomplished results and changes. However, metrics change with increasing process capability and experience. At the beginning, usually only simple, project-based metrics are collected, such as failure indices, effort indices, and schedule adherence. With increasing process capability, process-oriented metrics are added.[107]

---

106. The metric »functionality increase« (see figure 2–29) for project progress tracking may, for instance, also be used for trend statements, providing early warning if project objectives are jeopardized.

2.21 MAN.6 Measurement

Excursus: Goal/Question/Metric (GQM)Method

The central principle of the GQM method is that measurement is always goal-oriented. For this reason, measurement goals are defined based on an organization's business goals and substantiated by asking a number of questions. Based on these questions, metrics are derived which are supposed to provide the information to answer the questions. In answering the questions, measurement goals are operationalized, and it can be assessed whether or not the measurement objectives were reached. This is illustrated in figure 2–26.

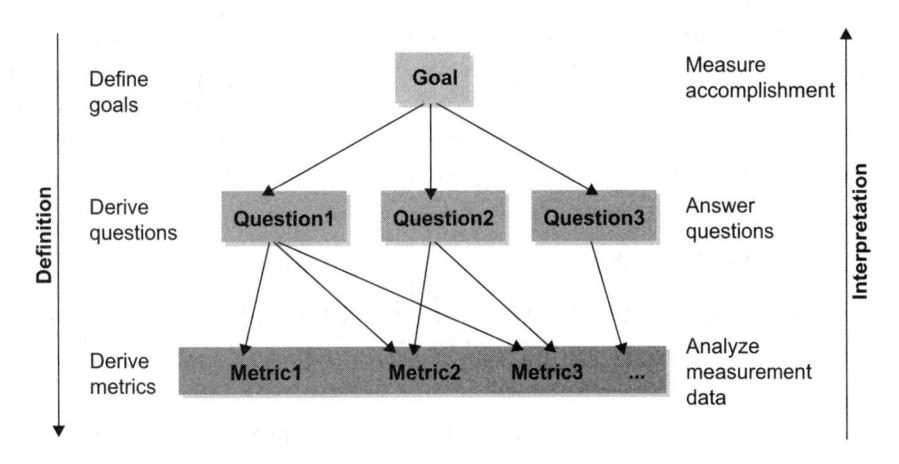

Figure 2–26   *GQM method*

To answer the questions, measurement data is collected and interpreted. It is important to ensure that measurement data is analyzed by people involved in the process (the data suppliers) or that the results of the analyses are at least made available to them. This ensures that the data is correct, accurately interpreted, and that the correct conclusions are drawn.

The questions are answered based on the measurement results, and ultimately the degree of goal achievement is measured. Measurements usually run through several cycles, and the goals, questions and metrics are refined in several steps (see [van Solingen et al. 1999]).

---

107. At Level 3, for instance, metrics are required that relate to the effectiveness and appropriateness of processes (see GP 3.1.5).

2.21.2 Characteristics Particular to the Automotive Industry

In the development departments of the automotive industry, measurement as a systematic process is the exception. Some possible reasons are:

- Compliance with production deadlines (Start of Production) and associated milestones, like product pre-series, sample delivery dates, etc., is given very high priority. Such being the case, a project's schedule adherence measurements, at first sight, typically yield 100 %. Only a closer look reveals that originally scheduled functionality has not been implemented in order to maintain schedule adherence.
- Fear of transparency and change: Capturing of effort per work package is difficult to implement in industrial practice. There are fears that conclusions can be drawn from data relating to the performance of individual members of staff. In Germany, for instance, the works council has the authority and obligation to co-determine decisions regarding the collection of such data.

2.21.3 Base Practices

BP1: Establish organizational commitment for measurement. *A commitment of management and staff to measurement is established and communicated to the organizational unit.*

Project control by means of measurement requires an open culture in the organization. Data suppliers, for instance, must have confidence that data they provide is used in a responsible manner. Open and sustainable management support for measurement is an indispensable prerequisite for a successful measurement process. Staff and management alike must be actively involved in the entire measurement process, and this must be clearly communicated to the organization.

BP2: Develop measurement strategy. *Define an appropriate measurement strategy to identify, perform and evaluate measurement activities and results, based on organizational and project needs.*

Based on the required strategy (often called »measurement concept«), the general measurement process must be developed and planned. This strategy is not to be considered independent from the various base practices of this process, but partly encompasses their results, too. Among others, it includes the following:

- How is the team structured that has operative responsibility for the implementation of this strategy? Which management representatives participate at the top level (e.g., in a steering committee)? How large is the available budget?
- Are there any restrictions/constraints to be observed regarding measurement (e.g., statutory provisions, involvement of the works council, etc.)?

2.21 MAN.6 Measurement

- How are the information needs of the organization identified (e.g., through interviews, working groups)?
- How are concrete metrics derived? What is the process of evaluation and approval by representatives of the organization (for instance, by performing multi-tiered reviews)?
- How can measurements be integrated in existing reporting channels (e.g., in project status reports, regular organizational meetings)?
- What needs to be done to promote acceptance and motivation of staff? In contrast, what are the »killer criteria«? How can fears regarding data abuse be allayed?
- What is the schedule for the activities to implement the process, which milestones are planned? Who is responsible for which activities?
- Who has overall responsibility for the measurement process? How is project progress regarding the first implementation of the process reported to management?

NOTE: *There are overlaps of this strategy with the generic practices of Level 2 for this process.*

BP3: Identify measurement information needs. *Identify the measurement information needs of organizational and management processes.*[108]

Information needs have to be elicited from the relevant individuals and groups involved in organizational processes (e.g., management tasks) and management processes (e.g., project management). This is typically done by interviews and/or working groups, in which executive managers, project staff, and other persons of the organization are included. If, for instance, the GQM method is used, measurement goals and questions are collected. Examples of information needs are:

- In the context of the supplier management process, the project manager wants to know how often a supplier shifts his deadlines, how many status meetings have actually taken place with the supplier, etc.
- To assess product quality, management, project manager, and quality manager are interested in quality indices like defect rates, or the defect detection efficiency of individual verification measures.

BP4: Specify measures. *Identify and develop an appropriate set of measures based on measurement information needs.*

Based on identified information needs and the measurement goals, metrics are now clearly specified. A clear specification particularly includes an accurate mathematical metric definition. A distinction must be made between basic met-

---

108. This does not exclusively refer to the organizational and management processes of ISO/IEC 15504 or Automotive SPICE. The terms are to be understood in a more general sense.

rics (which can be directly measured), and derived metrics resulting from mathematical operations using basic metrics. One example of a derived metric is the code review defect detection efficiency as the sum of all detected defects per KLOC (1000 lines of code). Measurement examples are:

- Required effort and time per work package
- Number of changes per unit of time
- Number of detected defects found during verification activities, related to the size of the measurement object
- Predicted residual defect density in products, etc.

It must also be clarified during metric specification who provides measurement values in which form, and at what time. It also makes sense to define control limits. For the »effort per work package« metric, for instance, the level of plan overrun can be specified at which intervention and analysis become necessary; also, what the causes are and which measures are to be initiated. It is also useful to have an effort estimate for the collection of the measurement values. Out of the very many possible metrics that are usually specified at the beginning, a reasonable selection must eventually be made. Figure 2–29 provides a sample specification of the »Functionality increase« metric.

BP5: Perform measurement activities. *Identify and perform measurement activities.*

The required activities are identified, planned, and performed according to the measurement strategy (BP2). They are described in base practices 6 to 11 in further detail.

BP6: Retrieve measurement data. *Collect and store data of both base and derived measures, including any context information necessary to verify, understand, or evaluate the data.*

Measurement data is identified, collected and stored. Besides measurement data, this also includes corresponding context information needed to evaluate the data correctly (e.g., When did the measurement take place? Where does the data come from? Where or in which process step was it collected?).

Raw data delivered by the data suppliers often require a plausibility check, and there may be queries and corrections. Measurement data should be correct, accurate, assigned to a specific activity or period, consistent and replicable [Fenton et al. 1997].

Data collection must follow a defined procedure (who, when, how, for whom). In practice, this is often done with the help of forms. If data is stored in too many different tables, files and documents, retrieval and prompt evaluation are difficult. It is advisable to use a database, or a few central documents that are stored in a specified location. They usually also support dedicated evaluations with role-related views on the data and evaluation results.

2.21 MAN.6 Measurement

Staff collecting the data must be trained accordingly and be informed about the measurement goals and the measurement process.

BP7: Analyze measures[109]. *Analyze and interpret measurement data and develop information products.*

Measurement data is analyzed and interpreted. Data providers should be involved in data evaluation or at least receive the results. The analysis is usually based on data that is presented in the form of easily understood graphs and tables. Figure 2–27 illustrates an example of a software product's »functionality increase«. In the example, project progress is tracked based on the implemented software functions. The analysis of the curve allows a prediction about the number of functions that will be implemented at the planned due-date (e.g., delivery). In the automotive industry, control of functionality increases is particularly important. Since schedule adherence usually has precedence over full functionality, planned functionality is often shifted to later deadlines if project progress is poor (e.g., implemented in a later development sample). In the worst case, it is not implemented at all.

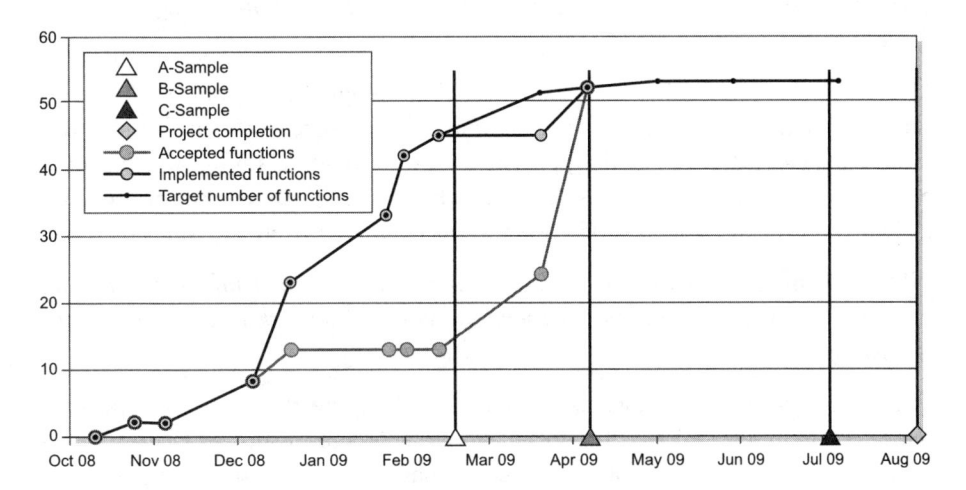

Figure 2–27   Example of a graphical evaluation regarding the »Functionality increase« metric

In a series of measurements with a large number of measurement values, statistical methods (consideration of data distribution, clusterings, scatter, etc.) can also be used. Typical analysis methods are:

- Analysis of measurement data in the project team, e.g., during normal project meetings, to derive control measures at project level

---

109. Automotive SPICE writes »measures«, although the correct term should be »measurement data«.

- Analysis of measurement data in steering committees or in regular meetings with management. There, the project manager may present and interpret evaluations that support management in its decision making
- Feedback meetings[110], in which data suppliers and the measurement team, or possibly process representatives, check the data and discuss and derive appropriate measures

Feedback meetings are a good instrument to verify and improve measurements. Example: A project manager wants to analyze how many defects are injected during which phases. Analysis of defect data gathered during tests shows that more than 50% of defects are classified as requirements related. At first glance, it looks as if those defects were injected during requirements analysis. Discussing the data in the team, however, soon brings to light that, by mistake, the 50% also includes errors injected during the design phase. The reason for this lies in inaccurate defect cause classification. As a result of the team discussion, the classification scheme is revised and explained to the team prior to the next measurement.

BP8: Use measurement information for decision-making. Make accurate and current measurement information accessible for any decision-making processes for which it is relevant.

Measurements are not an end in themselves but designed to support the business activities of the organization. They do this, for instance, by contributing to a more competent decision finding process and by providing stimuli for improvement measures. Based on the results of the analysis and, where appropriate, control limits as specified in BP4, decisions are made and concrete measures defined.

BP9: Communicate measures.[111]. Disseminate measurement information to all affected parties who will be using them and collect feedback to evaluate appropriateness for intended use.

The measurement evaluations as explained in BP7, and the resulting measures, must be communicated to all involved individuals and groups in the organization. These include data suppliers, project manager, executive managers, and process representatives. Ideally, measurement evaluations are an integral part of the existing reporting system and regularly (e.g., monthly) distributed. Suitable tools for disseminating measurement results at project level are project cockpit charts that combine the project's most relevant measurement values with traffic light evaluations [112] (see also MAN.3, Work Product 15-06—Project Status Report). Figure 2–28 provides an example.

---

110. A central element of the GQM method.
111. Automotive Spice uses the term »measures« by mistake.
112. Example: If a measurement result is within a certain area or interval, the light is on red. See also the exemplary specification of the measurement value Functionality Increase in figure 2–29.

2.21 MAN.6 Measurement

If measurement data are used intensively in the organization, it is always possible to assess if they fulfill their originally intended purpose. If they do, then metrics are, in our experience, continually developed further to satisfy the information needs of the user.

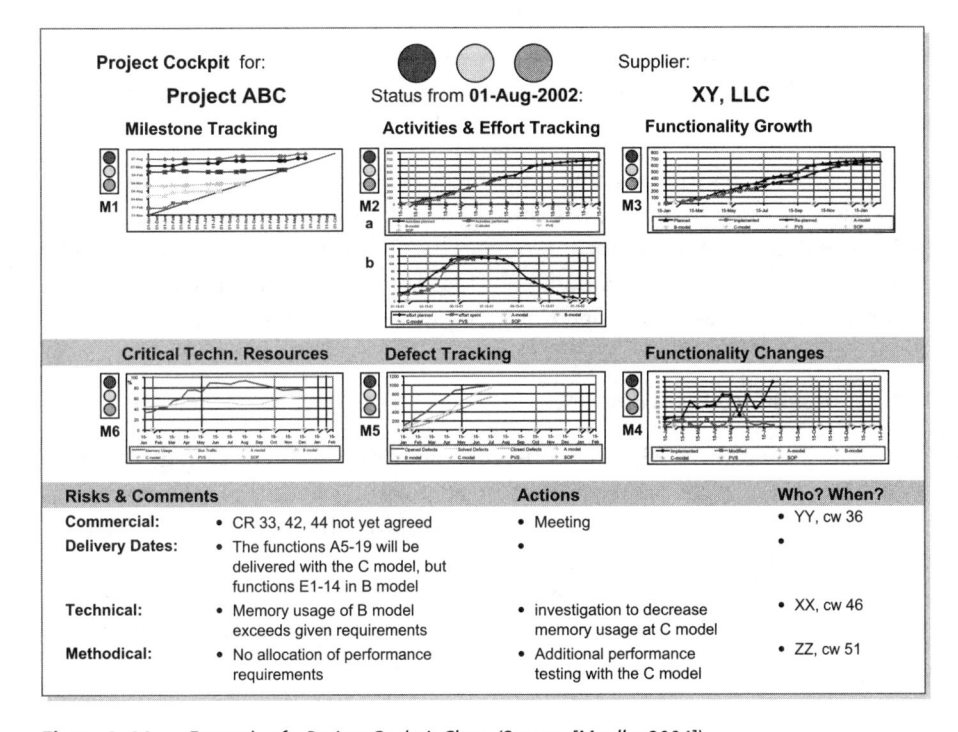

Figure 2-28 Example of a Project Cockpit Chart (Source: [Mueller 2004])

BP10: Evaluate information products and measurement activities. *Evaluate information products and measurement activities against the identified information needs and measurement strategy. Identify potential improvements.*

Experience shows that considerable time and improvements are often needed before measurement goals, metrics, and measurement results are stable and reliable enough to allow control of projects or processes based on this kind of information. The gained findings must therefore be repeatedly mirrored against the original objective. If necessary, objectives, measurement activities, and metrics must be adjusted to continuously improve the measurements.

Measurements can not only trigger improvements in the measurement process itself, but also in all other possible processes. The GQM method, for instance, distinguishes three different phases:

1. Understand: In the first phase, the development process and the work product characteristics become more transparent through measurement data. Process and product are better understood.

2. Control: If the measurement object is thoroughly understood, measurement data can be used to control the process or product creation by means of corrective and preventive measures.
3. Improve: Measurement data is now applied to identify improvement potentials and derive improvement measures.

BP11: Communicate potential improvements. *Communicate to the affected people the identified potential improvements concerning the processes they are involved in.*

Improvement suggestions must be communicated to the process representative of the process assessed by the measurement. He is typically the person who has to evaluate and implement these suggestions. This may affect the person responsible for the process at the organizational level, as well as the person responsible for performing the process at the operative level (e.g., in the project).

2.21.4 Designated Work Products

The work products 03-03 and 03-04 refer to data collected, stored, and processed as a result of measurements at project and organizational level. Data is available, for instance, in the form of project records (duration of activities, number of defects, etc.). In case of technical data (e.g., memory usage), it is available in the form of measurement records. In most cases data must be aggregated. For example, in order to analyze the number of found defects in reviews, data from individual review records is summarized in a table.

03-03 Benchmarking Data

The objective of benchmarking is to compare one's own performance with that of others or with historical data, and to derive improvement potentials. This presupposes a standardization of measurement values in order to be able to compare project data with data of other projects. Possible types of benchmarking are:

- Comparison within the organization: An OEM can, for instance, compare measurement values and product indices of a current development status with the last series production vehicle, or a supplier can compare product data with data of the predecessor product.
- Comparison with competitors
- Comparison of measurement values in the project team with historical data to derive control measures at project level
- Comparison of end-customer satisfaction (identified, for instance, by means of surveys and tests) with data from predecessor products

In the automotive industry, benchmarking is often used to allow comparison with competitors, for instance, regarding production times, production costs, number of defects or number of »break downs[113]«. OEMs maintain specific departments for competitor analysis to compare component parts and product quality. Internal benchmarking in electronics development, on the other hand, is rather uncommon. Possible benchmarking data would be ECU cost per item, defects after release, development effort, and development lead time. The capability profiles of ISO/IEC-15504 and Automotive SPICE can also be used as benchmarking data. The assessment scope must of course be comparable.

03-04 Customer Satisfaction Data

Customer satisfaction is usually measured subjectively through end-customer surveys (for instance, by means of standardized questionnaires) in the form of numerical values, e.g., a customer satisfaction index ranging from 1 to 5. Customer satisfaction can also be measured based on complaints and malfunctions in the field. Furthermore, there is also direct feedback from the customer which is perhaps documented in meeting minutes. However, such systematic feedback is fairly rare. If there is feedback, it is usually in the form of complaints.

> **Note to Assessors**
>
> Even though in Automotive SPICE there is no explicit work product for product-related measurement data, the collection of such data is also required depending on the information needs (BP3). For instance, product quality can be measured by the number of defects after delivery, the current number of known defects during development, and so on.

Various metrics are specified in the work products 07-01 to 07-08. They define indices that, for instance, describe the attributes of a procedure, a process, a product, or a target:

- Metrics for customer satisfaction
- Metrics for field measurements (e.g., what are the main causes for »break downs«)
- Metrics regarding processes at the organizational level (e.g., how many projects work in compliance with the prescribed processes)
- Metrics regarding processes at the project level, e.g., the functionality increase metric in project management)
- Quality metrics (e.g., number of defects detected by a specific type of test)
- Metrics regarding risks (e.g., number of highly prioritized risks currently defined in the project)

---

113. Term used for vehicles that become immobilized because of a malfunction and which must be towed to a car repair shop. This is the worst case from the customer's and manufacturer's point of view.

Figure 2-29 provides an example of a metric specification.

Functionality Increase
Objective: Track project progress by means of implemented software functions in order to be able to make trend statements
Associated Question: How does functionality increase in comparison to plan?
Description: Data collection: 1. Planned number of functions to be implemented at the reporting dates 2. Number of functions implemented at the reporting dates The implementation progress of functions is monitored to have an indication if the required functionality can be achieved at the reporting dates. Implemented functions are those that have been released by development. During the collection of implemented functions in connection with this metric, only those functions are taken into account that are required at the actual reporting date. (Other, additional and already implemented functions are to be taken into account during re-planning.) Depending on the project situation, additional, intermediate implementation states (e.g., coded, tested, accepted) may also be reported during the development process.
Activities at the Time of Reporting: ■ Update the actual curve at each software release (internal releases also count), not only at the sample phases ■ Compare with planning curve and determine deviations ■ Update the planning curve if there are new or modified requirements
Preconditions: An agreed function list (technically detailed view) with assigned completion dates is available. The function list is derived from the requirements specification and contains all the functions described in the requirements specification. The function list is adapted in case of requirements changes.
Data Definitions—Basic Measurements: ■ Planned number of functions for the reporting period (target number of functions) ■ Current number of functions implemented according to plan during the reporting period (accepted functions) ■ Number of functions newly planned for the following reporting periods ■ Total number of functions known at time of reporting
Data definitions—Derived Metrics: none
Data Source: Development report based on the function list
Chart: see figure 2-27 In addition to the accepted functions, the chart also shows all implemented functions (functions in status »implemented« that have not yet been formally accepted). ■ x-axis: Time line with reporting dates ■ y-axis: Absolute number of implemented functions or function groups Attributes for: target, actual, newly planned
Evaluation/Setting the Traffic Light Evaluation: ■ Green: If 95% or more of planned functions are completed at the time of reporting ■ Yellow: If less than 95% and 85%, or more, of planned functions are completed at the time of reporting. ■ Red: If less than 85% of planned functions are completed at the time of reporting

Figure 2-29 Metric specification for »Functionality Increase«

2.21.5 Characteristics of Level 2

On Performance Management

Much of the content listed in Process Attribute PA 2.1 is already required in BP2 as part of the measurement strategy. Activities are planned that are needed for the implementation of the measurement process, including schedule and responsibilities.

Measurement is usually specified as a standardized process on the organizational level. The measurement team defines measurement goals and metrics which are used in several projects. For this reason, the establishment of measurement as a process usually contains elements of level 3.

On Work Product Management

Particularly the measurement strategy, measurement values, and measurement evaluations should be reviewed.

2.22 PIM.3 Process Improvement

2.22.1 Purpose

The purpose of the Process improvement process is to continually improve the organization's effectiveness and efficiency through the processes used and aligned with the business need.

Good processes, good staff, and mastery of technology are admittedly the major influencing factors regarding costs, deadlines and quality. Processes play an important role, since they combine the other two factors to make up a powerful interactive system. Numerous world-class enterprises have impressively demonstrated how process improvement can lead to excellent commercial success, using improvement paradigms such as continuous process improvement (e.g., Toyota) and SixSigma (e.g., General Electric). Success through process improvement depends on two key factors:

- Consistently orientate processes towards the operational needs of the organization. In particular, this means using the processes systematically as tools to accomplish the business goals
- Undertake continued efforts, i.e., incorporate continuous process improvement as a part of the corporate culture and pursue it with vigor

2.22.2 Characteristics Particular to the Automotive Industry

The automotive industry is characterized by a high degree of effort relating to across-the-board process improvement. Above all, the reason for this is the decision of HIS to use SPICE and Automotive SPICE as tools for the evaluation of suppliers for electronics and software. Also, most of the OEMs are busy with corresponding programs (see also figure 1–3). First successes are already visible, particularly in enterprises that have undertaken massive efforts in this direction. Hampering factors include:

- Enormous time and cost pressure in the automotive industry
- Additional requirements, currently especially the IEC 61508 (see chapter 5), are to be implemented.
- Increasing project complexity due to an increased number of cooperation partners
- Steadily increasing internationalization and outsourcing
- Long project run times that tend to result in long process improvement implementation cycles
- Areas outside software development (e.g., hardware, mechanics) are often not involved in process improvement activities
- Problems at the organizational level are rarely addressed by the more project-oriented supplier assessments

2.22.3 Base Practices

BP1: Establish commitment. Commitment is established to support the process group, to provide resources and further abilities (trainings, methods, infrastructure, etc.) to sustain improvement actions.

NOTE: *The process improvement process is a generic process, which can be used at all levels (e.g., organizational level, process level, project level, etc.) and which can be used to improve all other processes.*

NOTE: *Commitment at all levels of management may support process improvement. Personal goals may be set for the relevant managers to enforce management commitment.*

There is a little joke among experts in process improvement:

Question: What are the three essential components of a successful improvement project?
Answer: 1. Management commitment
2. Management commitment
3. Management commitment

This humorous exchange reflects practical experience: Nothing works without solid management commitment! This applies to all management levels and is an effective means to include associated improvement objectives in the personal goals[114] of the managers in question. Especially at the beginning, much effort is put into a process improvement program to convince the different management levels and to establish lasting commitment. Once management is convinced, the improvement program should be endowed with appropriate priority and all necessary resources, and be actively and visibly supported by management. This can be done through participation in one of the program's steering committees. BP1 addresses three things towards which the commitment should be directed:

- Support of the process group: The process group is a group of process (improvement) experts, often called EPG or SEPG[115], constituting the core team for the process improvement program and for process maintenance. Often (but not necessarily), the manager of the group is also the manager of the process improvement program.
- Resources: Additional resources are necessary to ensure that the process improvement program is well integrated in the organization. This includes process experts who ensure that the processes are practical, as well as providing room for other members of staff to help in process improvement.
- Training, methods, infrastructure, etc.: Without training, coaching, and additional convincing activities, even the best process will never be applied in practice. Different methods (like an effort estimation model) and tools/infrastructure (e.g., for configuration management, change management, project management) are necessary to achieve a genuine benefit associated with the processes.

Of the levels addressed in the first notes (organizational level, process level, project level), the organizational level is particularly important since it supports the two other levels. Although it is possible to perform isolated process improvement for an individual process or project, its effectiveness and durability will remain rather limited without the support of the organizational level.

> **Note to Assessors**
>
> The process purpose clearly aims for organizational applicability; however, in its first note it states other possible areas of applicability (process, project). Planning the assessment, the process's scope must be defined early and prior to the interviews: Is it the project, is it a dedicated process, or is it the organization? Furthermore, the term *organization* must be clearly delineated. This is the only way that suitable interview partners can be invited in time.

---

114. This refers to the goals agreed between manager and senior management which are relevant for the manager's performance rating and the payout of variable allowances.
115. (Software) Engineering Process Group.

BP2: Identify issues. *Processes and interfaces are continuously analyzed to identify issues arising from the organization's internal/external environment as improvement opportunities, and with justified reasons for change. This includes issues and improvement suggestions addressed by the customer.*

NOTE: *Continuous analysis may include problem report trend analysis (see), analysis from Quality Assurance and Verification results and records (seeSUP.1 –SUP.2), validation results and records, and product quality measures like ppm and recalls.*

Basically, there are two ways to recognize improvement potentials:

- »Push principle«: Improvement suggestions are actively created by people involved within the organization. The organization must provide appropriate channels to receive such suggestions, for instance, in the form of an e-mail contact point[116], reward schemes for improvement proposals, etc.
- »Pull principle«: The organization undertakes its own efforts, for instance, in the form of internal assessments, employee surveys, evaluations of process or product-related metrics (e.g., from reviews, problem reports, or problems in the field).

Improvement proposals can refer to processes, interfaces between processes, organizational interfaces (internal as well as external), work product standards (such as document templates), tools, methods, and so on. Important hints often come from customers as a result of daily cooperation, and from supplier assessments.

BP3: Establish process improvement goals. *Analysis of the current status of the existing process is performed, focusing on those processes from which improvement stimuli arise, resulting in improvement objectives for the processes being established.*

Regarding analysis, also refer to the »Pull principle« in BP2. Focusing can be process oriented, for example on processes for which assessments identified weak points or where measurement data indicate the existence of problems. Frequent effort overruns in projects may be an indicator of weaknesses in the project planning and tracking processes. Focusing can also be more product oriented, using measurement data related to product quality and identifying the associated processes. Another alternative is the root cause analysis to identify the weak points in the process that caused a particular problem.

Improvement objectives can be qualitatively or quantitatively described, i.e., related to concrete, measurable targets. It is advisable, in each case, to agree on

---

116. Although not explicitly required in PIM.3, the organization is well advised to provide requesters with prompt feedback regarding the status of their request. It is our experience that otherwise the number of proposals will soon drop to zero.

2.22 PIM.3 Process Improvement

an accurate target definition that allows objective evaluation of the goal achievement.

BP4: Prioritize improvements. *The improvement objectives and improvement activities are prioritized.*

As a rule, many more improvement opportunities are identified than can be implemented with the available resources. In some cases, a particular sequence may be required because the individual improvement measures build on each other (e.g., improvement of requirements quality comes first, before measures related to design are initiated). For this reason, prioritization is indispensable. In doing so, the highest management level affected by these measures, as well as all involved managers of the levels below, should be involved to ensure their commitment. In the course of prioritization it is also advisable to point out the necessary effort needed for the improvement activities, (e.g., for process experts from the line organization, or pilot projects) and to secure the necessary resource commitments from the managers concerned. In the long run, it is best practice to establish a process steering committee or process CCB[117] for prioritization decisions.

BP5: Plan process changes. *Consequent changes to the process are defined and planned*

NOTE: *Process changes may only be possible if the complete supply chain improves (all relevant parties).*

NOTE: *Traditionally process changes are mostly applied to new projects. Within the automotive industry, changes could be implemented per project phase (e.g., productsample phases A, B, C), yielding to a higher improvement rate. Also, the principal of low hanging fruits (means implementing easy improvements first) may be considered when planning process changes.*

NOTE: *Improvements may be planned in continuous incremental small steps. Also, improvements are usually piloted before rollout at the organization.*

Once it is known which improvements are to be implemented, detailed planning becomes necessary. Which detailed changes are necessary with regard to processes, document templates, tools, etc.; which activities are required, who is responsible for this, and when are activities expected to occur? This is where the project planning for the improvement activities is done.

If process changes affect a complete supply chain, perhaps including an OEM, change processes naturally become sluggish. The choice of the scope of a change plays a decisive role regarding its chances for success. It may be better in some circumstances to start with a smaller scope, and to extend it after first, positive experiences have been gathered.

---

117. CCB = Change Control Board.

Whether running projects should be changed or not is an important decision that needs to be addressed in each individual case, taking effort and cost considerations into account. It is best to wait with this decision till after the pilot phase, and it is mandatory to include affected managers in this decision. The principle of the »low hanging fruits« can be extended to include a cost-benefit consideration and may already be applied during prioritization (see BP4).

According to the notes, the automotive industry already applies the principle of incremental development successfully in product development. Translated into process development, it means that processes should also be incrementally developed. Instead of aiming for the »perfect« process, experience has shown that it is better to gather experience with 80% solutions first before going on and then to improve them incrementally.

BP6: Implement process changes. The improvements to the processes are implemented. Process documentation is updated and people are trained.

NOTE: *This practice includes defining the processes and making sure these processes are applied. Process application can be supported by establishing policies, adequate process infrastructure (tools, templates, example artefacts, etc.), process training, process coaching and tailoring processes to local needs.*

Processes must be defined with the help of practitioners and, as we have already mentioned, be piloted prior to roll-out. In practice, the items listed in the note are essential for the processes' application. The policies mentioned ensure that members of staff are both motivated and committed in their efforts to apply these processes. Tailoring to local needs ensures that a generic process can be adapted, e.g., to suit different areas, departments, and project types. Incidentally, tailoring is a Level 3 requirement (see Generic Practice 3.2.1 in section 3.3.4).

BP7: Confirm process improvement. The effects of process implementation are monitored, measured and confirmed against the defined improvement goals.

NOTE: *Examples of measures may be metrics for goal achievement, process definition and process adherence*

The comparison of improvement progress with predefined goals opens up the opportunity to establish a continuous feedback cycle of progressive improvement consisting of setting objectives, implementation, and measurement of the goal achievement. Measurement of goal achievement presupposes that goals are measurably defined (see BP3). Appropriate measurements relating to the quality of the process definition are, for instance, employee surveys regarding process suitability. The measurement of process compliance can, for instance, be done by means of internal audits that measure the degree of compliance.

BP8: Communicate results of improvement. Knowledge gained from the improvements and progress of the improvement implementation is communicated outside of the improvement project across relevant parts of the organization and to the customer (as appropriate).

It is important to communicate successes and progress to the organization, also to strengthen the commitment and motivation of all involved parties. Suitable forms can be regular written reports to members of staff and oral reports at department meetings, works assemblies, and so on. Two factors positively facilitate the effect of this kind of communication:

- Improvements are based on figures, data, and facts that are actually verifiable.
- Improvements are reported by those involved in practical project work (e.g., project team member/s, manager), not by the process experts.

BP9: Evaluate the results of the improvement project. Evaluate the results of the improvement project to check whether the solution was successful and can be used elsewhere in the organization.

Subsequent to evaluations according to BP7, this evaluation helps to decide whether results are sufficient to begin extended rollout. The results may perhaps also be of benefit to other parts of the organization.

2.22.4 Designated Work Products

08-29 Improvement Plan

The improvement plan contains the improvement objectives derived from the organization's business goals, the organizational area of applicability, processes to be improved, appropriate milestones, designated reviews, reporting mechanisms, key roles, and responsibilities.

15-16 Improvement Opportunity

An improvement opportunity is described by the problem, the problem cause, proposals for problem resolution, the improvement's implementation benefit, and the damage caused by non-implementation of the improvement.

2.22.5 Characteristics of Level 1-3

On Level 1

Based on Base Practices 3 to 5, important planning mechanisms are already implemented in this process at Level 1.

On Level 2

At Level 2, planning and tracking has become highly perfected (Process Attribute PA 2.1). The requirements of Process Attribute PA 2.2 particularly apply to the improvement plan and to the numerous work products of the improvement teams. Initially, validation of these work products may be done via internal reviews performed by the process group, and later by means of user reviews. All work products are under configuration control.

On Level 3

The principal improvements for this process at Level 3 are the standardization of process improvement practices and the tailoring possibilities for different improvement teams. This is particularly interesting for larger organizations.

2.23 REU.2 Reuse Program Management

2.23.1 Purpose

The purpose of the Reuse program management process is to plan, establish, manage, control, and monitor an organization's reuse program and to systematically exploit reuse opportunities.

In times of increasing cost pressures and ever-shorter development cycles, the reuse of components [118] is becoming more and more important, raising high expectations for the benefit that may be achieved: improvements in reliability, performance, cost, and quality. This is in contrast to the additional costs (management effort, increased development and documentation effort, etc.) and risks (e.g., the broad distribution of defects) that need to be compensated for. Investing in a regular reuse program only pays off if the benefits significantly exceed the disadvantages. This is usually the case if there is a sufficient number of developers developing the same or at least similar products with sufficient reuse potential. Given these preconditions, organizations can expect substantial economic benefits.

2.23.2 Characteristics Particular to the Automotive Industry

The reuse of components has a long tradition in the automotive industry and has been common practice in many organizations for years[119]. In the area of software, for example, current development environments and various reference

---

118. REU.2 is written very generically. The term »component« can also be applied to hardware, mechanics, software, and system.
119. It is sufficiently known in the industry that this does not automatically lead to good quality, for instance, if older systems are successively upgraded with more and more features.

architectures support the development of software platforms, the derivation of variants, and the reuse of included software.

The most frequent difficulties for the reuse of software in the automotive industry are problems related to the portability of software that is very closely related to hardware. While coding for a specific microcontroller type, particular controller characteristics like the instruction set, assignment of memory space, or timing behavior, must be taken into account. Often only the tool suite belonging to the controller is fully covered. These tool suits are usually rather comprehensive and expensive and include, for instance, a special compiler, linker, debugger, emulator, and test functions. Software-porting to the next controller version takes a certain amount of effort even within one controller family; between different manufacturers it is almost impossible. To minimize this problem, many development organizations have engaged in strategic partnerships, typically with one controller manufacturer who, in return, provides improved support.

To minimize these disadvantages, a number of initiatives have been founded to support code reuse. In the automotive industry, the most important of these is AUTOSAR (see [AUTOSAR]; also refer to the glossary). The main objective of the AUTOSAR project is the definition of modular software architectures for automotive ECUs, with standardized interfaces and a standardized runtime environment. This way, software units of different origins (i.e., OEMs, suppliers) are expected to become exchangeable.

A different approach is model-based development including automatic code generation. Here, functionality is described at a higher abstraction level. After completion of the model, adaptation to the target can (at least theoretically) be done at the touch of a button.

2.23.3 Base Practices

BP1: Define organizational reuse strategy. *Define the reuse program and necessary supporting infrastructure for the organization.*

One refers to a reuse program if an organization reuses work products systematically and in a structured way. This includes enhanced requirements on work product quality and documentation, as well as a structured and user-friendly deployment of re-usable work products.[120]

First of all, the purpose, scope (or area of applicability), and goals for the reuse program must be defined. The purpose describes the economic motivation of the undertaking. In view of the rather considerable effort involved, it is advisable to try and achieve a broad consensus with the management of the organization. Regarding the scope, the following must be distinguished:

---

120. Improvised, ad hoc reuse of earlier project results does not constitute a reuse program!

- The organizational area of applicability (which parts of the organization are to be involved in the program?)
- Affected products or systems
- Types of work products to be reused

Reuse work products can be hardware components, mechanics components, system components, source code, parameter files, algorithms, interface definitions, design elements (e.g., UML descriptions), as well as requirements, processes, test cases, documentation, and knowledge (e.g., in the form of experience reports). In addition, the necessary infrastructure needs to be defined, including:

- Organizational structures, including staffing requirements
- Technical infrastructure regarding storage, administration, and provision of the reuse components, including all necessary software and hardware
- The method to be applied (processes and roles) regarding production, administration, maintenance, and provision of the reuse components
- Requirements on reuse candidates regarding content, quality, and documentation

BP2: Identify domains for potential reuse. *Identify set(s) of systems and their components in terms of common properties that can be organized into a collection of reusable assets that may be used to construct systems in the domain.*

A domain can be a product, a product family, or a product generation. Each domain embraces a large amount of already-developed systems, as well as running projects. It is important to identify which domains are potentially interesting to be included in the reuse program. A domain becomes interesting if its systems contain frequently recurring components which might be re-usable elsewhere. This investigation should be done together with experienced experts that have a good domain overview.

BP3: Assess domains for potential reuse. *Assess each domain to identify potential use and applications of reusable components and products.*

Each domain is now more closely inspected with regard to frequently recurring components, as already mentioned in BP2. These components must be analyzed to see how similar they really are in the different systems and whether a »common denominator« can be found. That is to say, is it conceivable that there is a generic component that can be adapted to suit different application purposes using hardware or mechanical variants, parameterizations, or adjustments in the code? In doing so, the effort required to provide the generic component and the total sum of adaptive work required for the different application purposes must be estimated to see whether it is really less than the sum of separate, individual development efforts. If possible, experience regarding effort in real projects should be taken into account. If no detailed effort records are available, an expert estimate should be obtained. This investigation can only be done by experienced domain experts (e.g., project manager).

BP4: Assess reuse maturity. *Gain an understanding of the reuse readiness and maturity of the organization, to provide a baseline and success criteria for reuse program management.*

On the one hand, reuse maturity, as described, plays a role in the decision whether or not to initiate a reuse program. On the other hand, if a positive decision has been made, recording of the reuse maturity at the time of the decision is essential for the verification of future improvements. For instance, what is the reuse quote after two years of development, how much effort could be saved through reuse (after deducting effort for deployment)? In the long run, a reuse program will only survive if such questions can be answered positively. The following factors contribute to reuse maturity:

- Reuse quote, i.e., the percentage of currently reused components
- Reuse potential, i.e., how many components could theoretically be additionally reused, as measured by the number of components developed per unit of time
- Factors that are currently opposed to reuse
- Factors that are currently in favor of reuse

BP5: Evaluate reuse proposals. *Evaluate suitability of the provided reusable components and product(s) to proposed use.*

This evaluation is done in two steps. First, a decision must be made regarding which of the components should be included in the reuse program.[121] The criteria for this selection are described in BP3. According to this practice, ideal components are those whose development is cheap, which can be used in many places with few adaptations, and which consequently save a lot of effort.

For each selected component, a decision must be made in a second step regarding which of its versions[122] is to be taken as the baseline for production.

BP6: Implement the reuse program. *Perform the defined activities identified in the reuse program.*

The reuse program is now being implemented. It is advisable to implement a preceding piloting phase to gather some experience with the reuse process. Based on these experiences, the program can then be extended step-by-step.

---

121. There are usually more proposals than can be implemented.
122. Since this component has been developed several times, a decision must be made whether to opt for a system A, B, or C component.

BP7: Get feedback from reuse. *Establish feedback, assessment, communication and notification mechanism to control the progress of reuse program.*

Feedback from »internal customers« (i.e., projects that use components out of the reuse kit or library) is important in order to be able to carry out adjustments to the program. If any discontent arises, the reuse program will be put at risk. Feedback can be given voluntarily (e.g., by directly addressing the person responsible for the reuse program, or through e-mails), or it can be actively obtained through interviews.

Evaluation mechanisms allow periodic (at least annual) comparisons with the previous baselines and success criteria. Periodic comparison is important to keep up economic motivation for the program and to adjust it when necessary.

Communication and reporting mechanisms inform the user of changes and alterations in the reuse portfolio, for example, regarding corrected defects and new releases. In case of newly detected critical defects, warnings need to be issued to the users.[123]

BP8: Monitor reuse. *Monitor the implementation of the reuse program periodically and evaluate its suitability to actual needs.*

Monitoring of the reuse programs is a task of management, which in BP1 became active promoter of the program. It does so, for instance, via regularly held steering committee meetings. Feedbacks obtained in BP7 can serve as a basis, especially the evaluation results addressed there. In view of the economic benefits like cost and time savings and quality improvements, this is an important management task.

2.23.4 Designated Work Products

08-17 Reuse Plan

The reuse plan contains the organization's management policies. In addition to the standards used for the development of reusable components, it also defines the reuse library or reuse kit. It also includes the list and description of the reusable components.

15-07 Reuse Evaluation Report

The reuse evaluation report contains possible reuse potentials, states all reuse related investments, and describes the current infrastructure. Above all, it contains the reuse program's current implementation status.

---

123. Especially in large reuse programs, it is rather counterproductive to burden a large number of an organization's internal users with reports that are not relevant to them. Target-oriented reports, however, require accurate accounting of who uses which component versions.

2.23.5 Characteristics of Levels 1-3

On Level 1

In this process, basic planning (BP1-4) and tracking mechanisms (BP7-8) are already implemented at Level 1.

On Level 2

At Level 2, planning and tracking become much more stringent and more detailed (Process Attribute PA 2.1). The requirements of Process Attribute PA 2.2 particularly apply to the reuse library and the reuse kit, as well as to the planning and evaluation documention. The review of the reuse library and reuse kit can be done by internal audits. Planning and evaluation documents can be reviewed jointly with management. The work products are under configuration control.

On Level 3

The essential innovation regarding this process at Level 3 is the standardization of the reuse process and the tailoring possibilities it offers, not only for different domains but also for different components.

2.24 Traceability in Automotive SPICE

2.24.1 Introduction

Starting with the requirements, traceability establishes a link between items of different development stages, for instance, between a requirement and associated test cases. This concept is also included in the ISO/IEC 15504 standard and in its previous version (the »Technical Report«). In Automotive SPICE it is considerably expanded and more consistently defined. In the ISO/IEC 15504 standard, the concept of traceability was flawed by numerous inconsistencies which were resolved in Automotive SPICE. Instead of »traceability«, Automotive SPICE uses the concept of »bilateral«, i.e., bidirectional traceability. This means that the link mentioned above can be tracked in both directions. In the meantime it appears that, unfortunately, the term »bilateral traceability« is often wrongly interpreted and applied.

2.24.2 Key-Notes

Two traceability variants can be distinguished in Automotive SPICE: horizontal and vertical traceability[124]. Using a V-model (see [V-model]) as a basis, vertical traceability is the traceability on the left side of the V-model from top to bottom and vice versa. Horizontal traceability is the traceability from the left side of the V-model to the right side at the same level, and vice versa (see figure 2–30 and figure 2–31).

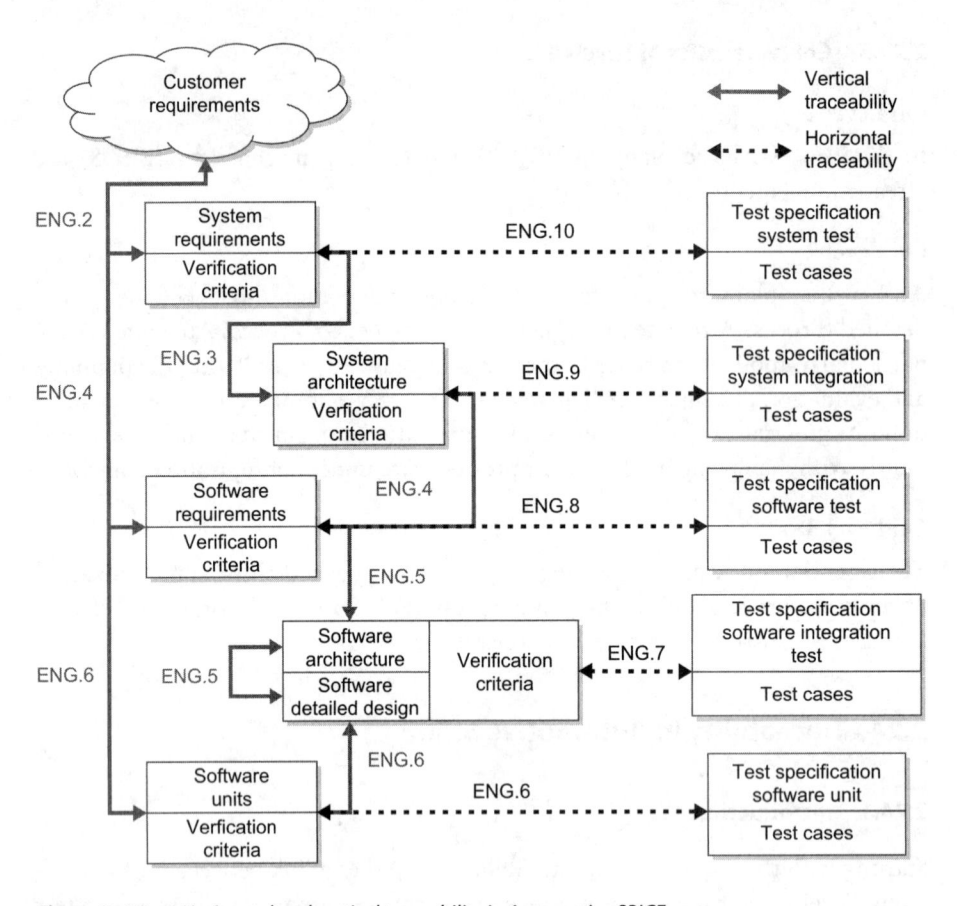

Figure 2–30 *Horizontal and vertical traceability in Automotive SPICE*

Bidirectional traceability can, for instance, be implemented with one or several traceability matrices. A traceability matrix is a table in which the relationships between items of different development phases are modeled, for instance, between requirements and associated test cases, and vice versa. Other implementations may be:

---

124. So far, the terms »vertical« and »horizontal« have not been mentioned in Automotive SPICE. However, they have been (and are currently still being) discussed and used in Automotive SIG and are to be integrated into the new traceability process. They serve as an illustration of traceability from a V-model perspective and differ from the CMMI perspective. In CMMI, vertical traceability is traceability along the development process, whereas horizontal traceability describes dependencies and cross-relationships between the components of a complex system.

2.24 Traceability in Automotive SPICE

- Referencing by name[125], i.e.,
  - Of code files in design documents, or identical naming of design elements and code files
  - Of design elements in the header of code files
  - Of requirement IDs in design elements
- Use of requirements management tools to record the relationships between requirements, design, code, and so on

> **Note to Assessors**
>
> In assessments, it is recommended to select requirement samples from the requirements specification and to track and evaluate their further development or implementation across the individual V-model processes.

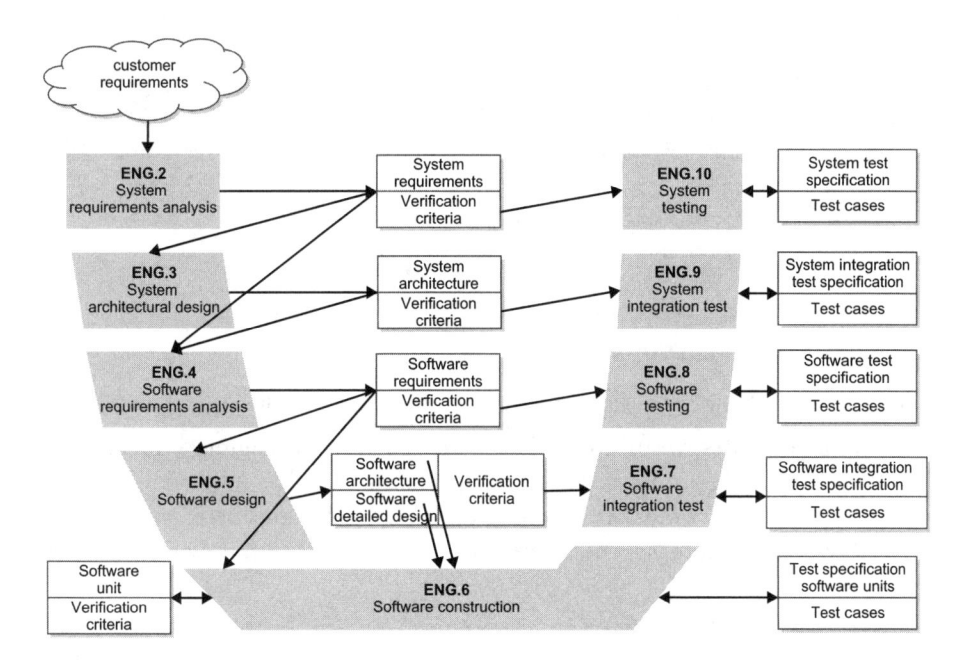

Figure 2–31 V-model type presentation of the engineering processes in Automotive SPICE and the input/output relationships of the work products

Benefits associated with bidirectional traceability include:

- Horizontal traceability can be used to verify test coverage easily.
- From the customer's perspective, vertical traceability can be used to verify that all requirements have been implemented.

---

125. Some organizations use internally developed programs to scan the files, using naming references to automatically generate traceability lists or matrices.

- If changes are necessary, vertical traceability from top to bottom and horizontal traceability from left to right can be used to quickly and completely detect all affected items to be changed, together with their impact on subsequent development steps. This way, scope and effort of the changes can be quickly identified. In the same way, all affected test specifications including test cases can be found quickly. This way, very efficient regression testing can be done for changes that have been implemented.
- Using horizontal traceability from right to left, and vertical traceability from bottom to top during verification and validation activities, the underlying requirements and design elements can be traced back very quickly. This allows very efficient verification whether or not there are faulty implementations or misinterpretations. The same applies to design and implementation activities, for which knowledge of corresponding requirements is very helpful.

> **Note to Assessors**
>
> For the implementation of bilateral traceability, tool support and a preferably consistent development environment are very helpful. However, if there are tool breaks in the development environment—particularly between OEM and supplier—and if therefore different traceability systems exist, particular attention must be paid to ensure that these are consistent.

Excursus: Verification Criteria

Automotive SPICE distinguishes between two different levels of verification criteria:

- In some processes, verification criteria are required for complete units, components, or work products. They specify what needs to be fulfilled for *a complete module, a component, or a work product* to be considered successfully verified (on this note, see the deliberations in SUP.2). The following processes are impacted:
  - In SUP.2 BP2 »verification criteria for work products« are developed. They define the needs to be satisfied for a *work product* to be considered successfully verified.
  - In ENG.6 BP2 verification criteria are requested for *complete software units*. There we also find content-related information regarding these verification criteria.
- In the engineering processes, the term »verification criteria« is used. These criteria describe what needs to be fulfilled before *a particular requirement* can be considered successfully verified.

2.24 Traceability in Automotive SPICE

In this excursus, we restrict ourselves to the term last mentioned. Each requirement listed in a requirement specification (WP-ID 17-00) must be verifiable, based on verification criteria. Verification criteria (WP-ID 17-50) have the following attributes:

- They are uniquely assignable to one requirement.
- The requirement is verifiable or can be evaluated.
- Verification criteria define the qualitative and quantitative criteria for the verification of a requirement.
- Verification criteria show that a requirement can be verified under agreed conditions.

Verification criteria, therefore, relate to requirements, i.e., to system requirements or software requirements. Verification criteria for the system architecture must relate to the system requirements and are therefore identical to the verification criteria for the system requirements. It is therefore enough to document them only once, thereby simplifying the traceability between system requirements and system architecture. This statement also applies to the relationship between software requirements and software architecture/software detailed design.

The verification criteria need to be specified during the development of the requirements; as such, they document criteria which are verifiable against the requirements. Figure 2–31 shows the input-output relationships of the affected processes and work products with consistent assignment of verification criteria to the different work products. The verification criteria must be established[126] early, i.e., simultaneously or close to the creation of the work products, rather than during the test processes in the right branch of the V-model. In many cases, verification criteria will need to be successively defined, for instance, if during the system architecture phase additional system requirements are detected[127] for which verification criteria need to be specified. The same applies to all the other processes in which system or software requirements are subsequently detected.

Verification criteria are an important, though not the only, input for the creation of the test plan, test design, test procedures, and test cases[128]. Regarding test, the following must also be considered:

- In ENG.3, architecture decisions that do not trace back to system requirements (and for which originally no verification criteria existed), need to be considered in system integration testing.
- In ENG.6, the verification criteria mentioned for the respective software unit in BP2 need to be considered in unit testing.

---

126. This requirement is a result of the base practices requiring traceability on the V-model's left side. The work product characteristics of the work products created there also require verification criteria.
127. In this case they need to be updated in the system requirements.
128. See the excursus on »Test Documentation« in ENG.6.

Examples of Verification Criteria for System Requirements

Requirement 1: Corrupted content can be found in the EEPROM.

Verification Criterion 1: Corrupted content in the EEPROM can be found by a check sum calculation of the diagnostic service »readCheckSumStatus«.

Requirement 2: If the content in the EEPROM has been corrupted, default values shall be used for all variables initialized by the data in the EEPROM.

Verification Criterion 2: All variables that are initialized by data in the EEPROM can be read via the diagnostic service »readDataByAbsoluteAddress«.

The following paragraphs provide an overview of the content-related aspects whereby vertical traceability relationships[129] are considered first:

- Customer requirements—System requirements: For each customer requirement the derived system requirements are known, and vice versa: for each system requirement the sources are known.
- System requirements—System architecture: For each system requirement it is known in which system architecture element it is implemented. For each system architecture element the relevant system requirements are known. Some system requirements can perhaps not be mapped onto system architecture elements. This particularly applies to nonfunctional system requirements.
- System requirements—Software requirements: For each system requirement the derived software requirements are known, and vice versa: for each software requirement the system requirements are known. Some system requirements can perhaps not be mapped onto the software requirements. This particularly applies to non-software relevant system requirements.
- System architecture—Software requirements: For each system architecture element the relevant software requirements are known, and vice versa: for each software requirement the relevant system architecture elements are known.[130] Some system architecture elements can perhaps not be mapped onto software requirements, this particularly applies to non-software relevant system architecture elements. The verification criteria regarding traceability can be ignored, as they are already covered by the traceability between system requirements and software requirements.
- Software requirements—Software architecture: It is known for each software requirement in which software architecture element it is going to be implemented. For each software architecture element the relevant software require-

---

129. To make the text more readable, verification criteria included in the respective work products are not explicitly mentioned.
130. Either is easy to implement if the structure of the software requirements document reflects the system architecture.

ments are known[131]. Some software requirements can perhaps not be mapped to software architecture elements. This particularly applies to nonfunctional software requirements. If verification criteria are already exhaustively described in the software requirements, they may be disregarded here as far as traceability is concerned.

- Software architecture—Software detailed design: It is known for each software architecture element in which element of the software detailed design it it is going to be implemented. For each element of the software detailed design the relevant software architecture elements are known[132]. If verification criteria are already exhaustively described in the software requirements, they may be disregarded here as far as traceability is concerned.
- Software requirements—Software units: It is known for each software requirement in which software unit it is going to be implemented. For each software unit the relevant software requirements are known. Some software requirements can perhaps not be mapped to software units. This particularly applies to nonfunctional software requirements. If verification criteria are already exhaustively described in the software requirements, they may be disregarded here concerning traceability.
- Software units—Software detailed design: It is known for each software unit in which element of the software detailed design it is going to be implemented. For each element of the software detailed design the relevant software units are known[133]. If verification criteria are already exhaustively described in the software requirements, they may be disregarded here concerning traceability.

The following list provides an overview of the content-related aspects of horizontal traceability relationships:

- Software units including verification criteria—Test specifications of the software units including test cases: For each software unit, the associated test specification and the test cases must be known, and vice versa. For the verification criteria[134] the test cases must be known, and vice versa.
- Software architecture and software detailed design—Test specifications of the software integration test, including test cases: For each element of the soft-

---

131. Either is easy to implement if the structure of the software requirements document reflects the software architecture.
132. Either is easy to implement if software architecture elements, software components, and software units are explicitly described in the form of a hierarchical decomposition.
133. Either is easy to implement if each software unit can be uniquely referenced to a software detailed design element of the same name, for instance, in the form of a separate file for the software unit detailed design or in the form of a chapter in a higher-level document.
134. These verification criteria are identical to those of the software requirements corresponding to the software unit.

ware design relevant for integration testing (typically interfaces, cross-module functions), the associated test specification and the test cases must be known, and vice versa.
- Software requirements—Test specifications of the software test, including test cases: For each software requirement the associated test specification and the test cases must be known, and vice versa.
- System architecture—Test specifications of the system integration tests, including test cases: For each element of the system architecture relevant for integration testing (typically interfaces, cross-module functions), the associated test specification and the test cases must be known, and vice versa.
- System requirements—Test specifications of the system test, including test cases: For each system requirement the associated test specification and the test cases must be known, and vice versa.

Note to Assessors

In the end, the granularity of the traceability depends on the concrete project. In any case, it should be warranted that all requirements can be traced quickly down to the source code and from test cases back to the requirements.

Not all traceability relationships must be explicitly documented. For instance, in development processes with auto-code generation, the tracking of software requirements to software units can be made indirectly, tracking the software requirements to the software design (in the auto-code tool) and from there to the source code per code generation. A final evaluation of what needs to be available and what not is left to the judgement of the assessor.

3 Interpreting the Capability Dimension

This chapter consists of three sections: In sections 3.1 and 3.2 we explain the structure of the capability dimension and the evaluation methodology.

While the process dimension indicators are process-specific (see chapter 2), the process attributes and the indicators for the capability dimension are generic, i.e., they are to be applied to all the processes of the process dimension. This is why they need to be interpreted in the context of each individual process. Section 3.3 discusses the capability levels and associated practices in detail and provides general interpretational support. Additional process-specific interpretational support is given in chapter 2 in the respective processes. We will show in detail what is needed to achieve Levels 1–3. Levels 4 and 5 will only briefly be touched upon, since many organizations focus on Levels 1–3.

3.1 The Structure of the Capability Dimension

3.1.1 Capability Levels and Process Attributes

The capability dimension describes the maturity of processes using nine process attributes assigned to the Levels 1–5. Level 0 does not have a process attribute assigned. The process attributes are defined in the normative measurement framework (see ISO/IEC 15504-2). Section 4.2.3 of the ISO/IEC 15504-5 standard defines the term »process attribute« as follows:

Process attributes are features of a process that can be evaluated on a scale of achievement, providing a measure of the capability of the process. They are applicable to all processes...

Each capability level is assigned one (at Level 1) or two (from Level 2 onward) process attributes. Each process attribute describes a particular aspect of the respective capability level. All the requirements of the capability levels build upon each other. Level 3, for instance, includes all the requirements of the lower Levels 1 and 2.

3.1.2 Process Capability Indicators

In the assessment model, indicators are used to describe the process attributes in more detail. These indicators provide assessors with interpretative guidance to help them in their rating of the process attributes.

The indicators of the capability dimension describe key activities, resources or outcomes related to the process attributes. In Automotive SPICE, there are two kinds of indicators for the capability dimension:

- Generic Practices (GP)
- Generic Resources (GR)

Generic practices are generically formulated activities which provide guidance for the implementation of the respective process attributes. Many of the generic practices support the performance of the base practices through process management activities. One example of a generic practice is GP 2.1.2 (Plan and monitor the performance of the process), which requires the implementation of basic project management principles for the respective process; i.e., that the process activities are scheduled and their performance is tracked in accordance with the plan.

To perform the generic practices, generic resources can be used. Generic resources include staff, tools, methods, or other infrastructure. For instance, a method to perform reviews on work products that support Process Attribute PA 2.2 is one example of a generic resource.

In practice, we primarily use generic practices as indicators for process evaluation. The generic resources will therefore no longer be considered here.

To assess Level 1, indicators for the capability dimension (base practices and work products) are relevant because they are used to assess Process Attribute PA 1.1, and hence the performance of the process. Since these indicators are process-specific, they are described for each individual process in chapter 2 of this book.

3.2 How Are Capability Levels Measured?

During an assessment, the performed process is compared with the indicators of the assessment model. Based on this, the achievement of the process attributes is assessed. To determine the capability level of a process, ISO/IEC 15504-2 requires the evaluation of the achievement of the process attributes by means of the following 4-stage rating scale:

- N (Not achieved), corresponds to 0–15 %.
  There is little or no evidence of achievement of the defined attribute in the assessed process.
- P (Partially achieved), corresponds to >15–50 %.
  There is some evidence of an approach to, and some achievement of, the defined attribute in the assessed process. Some aspects of achievement of the attribute may be unpredictable.

- L (Largely achieved), corresponds to >50–85 %.
  There is evidence of a systematic approach to, and significant achievement of, the defined attribute in the assessed process. Some weakness related to this attribute may exist in the assessed process.
- F (Fully achieved), corresponds to >85–100 %.
  There is evidence of a complete and systematic approach to the defined attribute in the assessed process. No significant weaknesses related to this attribute exist in the assessed process.

> **Notes for Assessors**
>
> Although Automotive SPICE only requires a rating of the process attributes, base practices and generic practices are usually also rated using the N/P/L/F scale. This scale is used because it supports objective and reproducible rating very well. Work product quality and availability are also considered during the rating of the base practices, since work products are created and developed by means of the base practices.
>
> Many assessment methods calculate a mean value of the different base practice ratings as an indicator for the evaluation of PA 1.1, as well as a mean value of the ratings of the generic practices for the rating of the other process attributes. Assessment tools even calculate a percentage achievement of the process attributes from the mean value of the percentage achievements of the individual base practices or generic practices. However, this method can only serve as a guiding principle, since the base practices do not usually have the same weight, and may have a different weighting in the individual situation. The same applies to the generic practices.
>
> This problem plays a particular role in borderline cases where, for instance, an assessment tool proposes an achievement degree of 50 % (=P) and where the question is raised whether it should not be 51 % (=L) after all. Especially in such borderline cases the assessment team must base its assessment of the achievement of the process attributes on the verbal characterization of the N/P/L/F scale (see above) within the respective context, rather than on a percentage figure generated by some assessment tool. Another thing that needs to be considered is that the generic practices are by no means a precise mirror image of the characteristics of the process attributes. In fact, they are much more detailed. Some details (that may have been assessed with the generic practice) do not apply at the higher level of the process attributes. The same applies to the special case of PA 1.1, the evaluation of which aims at the process outcomes. In analogy to our remarks above, base practices are also no accurate mirror image of the process outcomes.

To calculate the capability level from the ratings of the individual process attributes, ISO/IEC 15504-2 provides a »process capability level model«. According to this model, the following applies:

- To achieve a particular level, the PAs of that capability level must at least be rated L (»Largely achieved«)
- All PAs of the lower levels must be rated F (»Fully achieved«).

This calculation is shown in figure 3–1 (see also ISO/IEC 15504-2, table 1).

Capability Levels	Process Attributes	Rating
Level 1	PA 1.1	Largely or fully achieved
Level 2	PA 1.1	Fully achieved
	PA 2.1	Largely or fully achieved
	PA 2.2	Largely or fully achieved
Level 3	PA 1.1	Fully achieved
	PA 2.1	Fully achieved
	PA 2.2	Fully achieved
	PA 3.1	Largely or fully achieved
	PA 3.2	Largely or fully achieved
Level 4	PA 1.1	Fully achieved
	PA 2.1	Fully achieved
	PA 2.2	Fully achieved
	PA 3.1	Fully achieved
	PA 3.2	Fully achieved
	PA 4.1	Largely or fully achieved
	PA 4.2	Largely or fully achieved
Level 5	PA 1.1	Fully achieved
	PA 2.1	Fully achieved
	PA 2.2	Fully achieved
	PA 3.1	Fully achieved
	PA 3.2	Fully achieved
	PA 4.1	Fully achieved
	PA 4.2	Fully achieved
	PA 5.1	Largely or fully achieved
	PA 5.2	Largely or fully achieved

Figure 3–1 Determination of levels using the »Process capability level model«

3.3 The Capability Levels

3.3.1 Level 0 (»Incomplete Process«)

The process is not implemented, or fails to achieve its process purpose. At this level, there is little or no evidence of any systematic achievement of the process purpose.

Level 0 is the only capability level that does not have a process attribute. Level 0 is awarded if the Level 1 Process Attribute PA 1.1 is rated P »Partially achieved« or N »Not achieved«.

3.3.2 Level 1 (»Performed Process«)

The implemented process achieves its process purpose.

»The purpose of the process is achieved« means that the process outcomes have been achieved.

In contrast to the higher levels, Level 1 has only one process attribute, PA 1.1 »Process performance attribute«, and one generic practice GP 1.1.1 »Achieve the process outcomes«. This process attribute refers to the purpose and process outcomes of the individual process; and these outcomes vary from process to process.

Process Performance Attribute (PA 1.1)

The process performance attribute is a measure of the extent to which the process purpose is achieved. As a result of full achievement of this attribute:

a) *the process achieves its defined outcomes.*

The achievement of the process purpose is measured by the extent to which the process achieves its outcomes. This evaluation is based on the indicator GP 1.1.1 (Achieve the process outcomes) which in turn is based on the indicators for the process dimension[1] (base practices and work products). Thus, base practices and work products are indicators of the extent to which process purpose and process outcomes are achieved. Figure 3–2 shows the relationships:

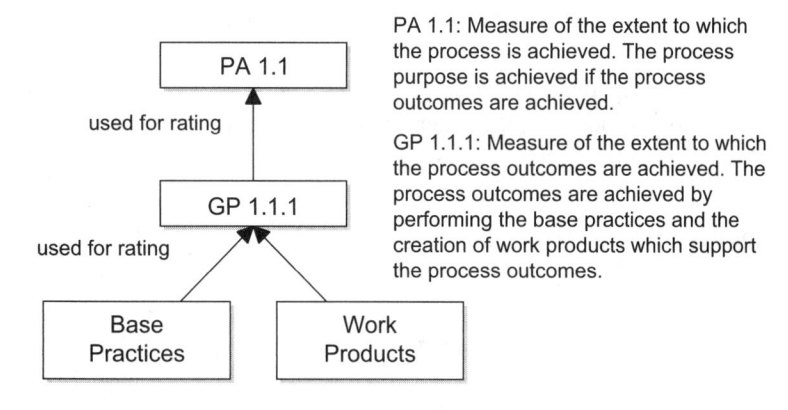

Figure 3–2   *Assessment of PA 1.1*

---

1. The indicators for the process dimension are only addressed by GP 1.1.1 and PA 1.1, and not by the higher level process attributes.

> **Notes for Assessors**
>
> The problem with the rating of PA 1.1 is that the process outcomes and the base practices are similar in content but not identical. The thematic relationship is established through referencing from the base practices to the process outcomes. However, the base practices are rather more detailed and sometimes accord different emphasis, or they differ in content slightly from the process outcomes.
>
> In most assessments these differences are negligible. They only matter if one particular process attribute rating is on the rocks and if the achievement of a level depends on it (i.e., in case of an L or F rating). In such cases we recommend consulting the process outcomes since they are of superior relevance in the rating hierarchy compared with the base practices and work products (see figure 3–2).

*GP 1.1.1: **Achieve the process outcomes**. Perform the intent of the base practices. Produce work products that evidence the process outcomes.*

The base practices are performed/complied with and appropriately implemented, i.e., they contribute to the process purpose and process outcomes and produce the work products.

The work products are appropriately implemented following the requirements of the assessment model (see Automotive SPICE PAM, Annex B »Work product characteristics«). The work products are distributed using suitable mechanisms. Work products are provided to users that need them.

It is not necessary for all work products of the process to be implemented, although the necessary work products[2] must of course exist and their technical content must be sound.

On the whole, all process outcomes should be achieved through the base practices and work products. Since there is only one generic practice, the evaluation of PA 1.1 depends entirely on GP 1.1.1.

> **Notes for Assessors**
>
> For the purpose of illustration, figure 3–3 lists some schematic rating examples. Cases 1, 2 and 4 do not contain any special characteristics and are not explained further. In case 3, sufficient test planning and corresponding test cases were available. The software units were coded according to coding guidelines. However, developer tests were insufficiently performed and documented because delivery dates were too aggressive. During the assessment, it was very difficult to reconstruct which tests had actually been executed. Because of this, the process attribute was rated P (i.e., Level 0: The process is not implemented, or fails to achieve its process purpose), although 6 of 7 base practices were rated F.

---

2. The assessor must decide in each individual context which work products are necessary.

3.3 The Capability Levels

Sample Process: Software Construction	Case 1	Case 2	Case 3	Case 4
Rating PA.1.1 Process Performance	P	F	P	L
O1 A unit verification strategy is defined				
O2 Software units defined by the software design are produced				
O3 Consistency and bilateral traceability are established ...				
O4 Verification of software units ...				
O5 Results of the unit verification are recorded				
ENG.6.BP1 Define a unit verification strategy	L	F	F	L
ENG.6.BP2 Develop unit verification criteria	L	F	F	F
ENG.6.BP3 Develop software units	L	L	F	L
ENG.6.BP4 Verify software units	L	F	N	F
ENG.6.BP5 Record the results of unit verification	L	F	N	F
ENG.6.BP6 Ensure ... traceability of SW det. design to SW units	P	F	F	L
ENG.6.BP7 Ensure ... traceability of SW req. to SW units	P	L	F	P
ENG.6.BP8 Ensure ... traceability of SW units to test spec	P	F	F	L
WP 08-52 Test plan		X	X	(X)
WP 08-50 Test specification	X	X	X	X
WP 11-05 Software unit	X	X	X	(X)
WP 13-22 Traceability record		X	X	(X)
WP 13-50 Test result	X	X	(X)	X
Legend: X :Work product available in the required quality (X):Work product available with significant deficiencies BP:Base Practice WP:Output work products (selection) O:Process outcome				

Figure 3–3   Rating examples for PA 1.1

3.3.3 Level 2 (»Managed Process«)

The previously described performed process is now implemented in a managed fashion (planned, monitored and adjusted) and its work products are appropriately established, controlled and maintained.

Level 2 means that in addition to the Level 1 requirements, process performance is now consistently planned and tracked. This increases the predictability of the achievement of work results and schedules. Furthermore, the process's work products are adequately monitored and maintained. One can therefore be more certain that the process work products meet the requirements.

Automotive SPICE assumes the existence of necessary work products. This does by no means imply that all work products listed in Automotive SPICE need to be available for each process. Whatever is considered »appropriate« must be judged within the context of the project or organization.

At Level 1, focus is on work products that directly belong to the base practices (e.g., design documents, code, test documentation). During transition to Level 2, generic process work products (e.g, review records, planning data related to process activities, and reports) and generic practices are added, both of which need to be managed.

Automotive SPICE requires documented processes explicitly only from Level 3 onwards. At Level 2 it requires, among other things, that the scope of work is determined for the performance of the process (GP 2.1.1) and that the process activities are defined (GP 2.1.2). This requirement is facilitated by a process description. In order to facilitate the planning of more complex processes, or to make planning sometimes possible at all, it is urgently recommended to provide a description of the process already at Level 2. In these cases, the documentation does not need to meet all the requirements of a Level 3 process description (see GP 3.1.1).

> **Notes for Assessors**
>
> It is highly unlikely for anyone to be able to plan and track a complex workflow without a process description. The decision as to what extent a missing process description is taken into account in the rating of PA 2.1 is left to the assessors.

Figure 3–4 provides an overview of the Level 2 process attributes and generic practices.

3.3 The Capability Levels

Process attributes		Generic practices	
PA.2.1	Performance management	GP 2.1.1	Identify the objectives for the performance of the process.
		GP 2.1.2	Plan and monitor the performance of the process to fulfill the identified objectives.
		GP 2.1.3	Adjust the performance of the process.
		GP 2.1.4	Define responsibilities and authorities for performing the process.
		GP 2.1.5	Identify and make available resources to perform the process according to plan.
		GP 2.1.6	Manage the interfaces between involved parties.
PA.2.2	Work product management	GP 2.2.1	Define the requirements for the work products.
		GP 2.2.2	Define the requirements for documentation and control of the work products.
		GP 2.2.3	Identify, document and control the work products.
		GP 2.2.4	Review and adjust work products to meet the defined requirements.

Figure 3–4   *Level 2 process attributes and generic practices*

PA 2.1 Performance Management Attribute

The performance management attribute is a measure of the extent to which the performance of the process is managed. As a result of full achievement of this attribute:

a) objectives for the performance of the process are identified;
b) performance of the process is planned and monitored;
c) performance of the process is adjusted to meet plans;
d) responsibilities and authorities for performing the process are defined, assigned and communicated;
e) resources and information necessary for performing the process are identified, made available, allocated and used;
f) interfaces between the involved parties are managed to ensure both effective communication and also clear assignment of responsibility.

The generic practices (GP) of PA 2.1 describe fundamental project management principles, yet applied to the respective process (and not to the project as a whole).

They require the process to be planned based on objectives, and that compliance to the plan is tracked. This includes the allocation and use of resources and the control of interfaces to involved individuals and groups.

Process planning and tracking according to PA 2.1 are particularly supported by practices of the project management processes (MAN.3).

Planning and tracking intensity may vary from process to process, and both activities can be done in different ways. It does therefore not make much sense to plan certain process activities at all. For instance, during the planning of the configuration management process (SUP. 8), it would not be practicable to plan and track the checking-in of every module into the CM tool as a separate activity. The build of the release, however, must be planned.

GP 2.1.1: Identify the objectives for the performance of the process.

NOTE: *Performance objectives may include:*
1. *quality of the artefacts produced*
2. *process cycle time or frequency*
3. *resource usage*
4. *boundaries of the process*

Performance objectives are identified based on process requirements. The scope of the process performance is defined. Assumptions and constraints are considered when identifying the performance objectives.

Targets for the performance of the respective process are set, for example:

- Quality of the process work products (e.g., residual defect density, achievement of requirements such as the correct use of a template, or the performance of a review)
- Process cycle times (e.g., how long is a particular process activity supposed to last?)
- Relevant due-dates/milestones
- Effort targets, based on effort estimates or experience
- Use of specific, qualified resources (minimum qualification, process experience/training)
- Which resources are used, and to what extent?
- Which activities are covered by the process, and which are not?

These targets may be either qualitative (e.g., »peer-review results must be documented«), or quantitative (e.g., »at least 80% of the work products are verified by peer reviews«).

The scope of work of the process must also be identified, i.e., the activities required within this process must be determined.

3.3 The Capability Levels

> **Notes for Assessors**
>
> For most users, the consideration of process objectives during the planning of a process is not a separate, conscious activity. In our experience, it makes sense to have the planning process explained first and then to ask for the project objectives and requirements. Requirements often result from many different documents (e.g., project goals, quality plan, project schedule, defined standard processes of the organization). An often encountered weakness is that the effort for cross-sectional processes (e.g., configuration management, project management) is not known. Although someone has been assigned to those tasks, they have also many other tasks and it is doubtful if due to work overload all required process activities are sufficiently performed.
> »Process performance« is often confused with »capability of a process«. In assessments, chances are that one will (rightly) encounter blank faces when asking the question »What are your process capability objectives?«. »Process performance objectives« would be the correct term.

> **Notes for Assessors**
>
> The following questions may help in interviews to identify the implementation of this GP:
>
> - What is the scope of work for this process?
> - Which requirements must be considered during the planning of these tasks, for instance, activity duration, estimated effort, quality of the work products?

GP 2.1.2: Plan and monitor the performance of the process to fulfill the identified objectives. Plan(s) for the performance of the process are developed. The process performance cycle is defined. Key milestones for the performance of the process are established. Estimates for process performance attributes are determined and maintained. Process activities and tasks are defined. Schedule is defined and aligned with the approach to performing the process. Process work product reviews are planned. The process is performed according to the plan(s). Process performance is monitored to ensure planned results are achieved.

GP 2.1.2 describes how principles of project management are applied to a process. »Process performance cycle« refers to procedures for process performance, as often specified by a process definition. »Process performance attributes«, for instance, can refer to size or complexity of work products but particularly refer to the effort required for process activities. Defined process activities constitute the basis for planning (on this note, also refer to the deliberations regarding documented processes on page 220).

GP 2.1.2 should not be confused with project management for the project as a whole. For example, »to plan the process activities« for the project management process (MAN.3) means that project management activities (such as project planning, creation of status reports) are planned. Project planning, on the other hand, comprises the planning of all project activities[3] and is required by the base practices of MAN.3. There is an overlap between GP 2.1.2 and MAN.3 regarding project-related processes[4], since their process activities are also project activities. The realization of GP 2.1.2 can thus easily happen within the context of project management, using the same planning documents. It is therefore not necessary to create a separate process planning document for each process. Planning can, for instance, be combined in one project plan. All reasonably plannable process activities should at least be planned and documented with dates, allocated resources, and responsibilities.

In some processes, it makes little sense to schedule individual activities because they are triggered by asynchronous events. Planning for regularly recurring activities (e.g., team meetings) is usually not done in the project plan; instead, applications like MS Outlook or other scheduling systems are used. However, effort estimation is mandatory in both cases, which means that recurring activities must also be considered.

If activities are repeated during the entire course of the project, for instance, configuration management, it may be difficult to accurately estimate the necessary staff requirements. It is therefore recommended to resort to the experience of previous projects with a similar project context and similar tasks.

Notes for Assessors

The following questions may help in interviews to identify the implementation of this GP:

- How do you plan the activities of the process? Are there agreed dates and milestones?
- How did you identify the complexity of the tasks and the associated effort?
- Did you plan reviews of the work products?
- How do you monitor process performance?

GP 2.1.3: Adjust the performance of the process. Process performance issues are identified. Appropriate actions are taken when planned results and objectives are not achieved. The plan(s) are adjusted, as necessary. Rescheduling is performed as necessary.

In case of deviations from the plan, appropriate measures must be initiated. Possible measures are:

---

3. All activities performed within the context of the project.
4. This refers to all processes performed in a project, as opposed to processes which are not implemented in projects, for instance, the process improvement process PIM.3.

- Corrective measures, in order to get the performance back in line with the plan
- Adjustments to the plan if the plan turns out to be no longer feasible. This may lead to comprehensive revisions of the plan.

If new planning or substantial replanning becomes necessary, a new agreement should be negotiated with the stakeholders. This agreement should be reviewed and officially approved.

> **Notes for Assessors**
>
> The following questions may help in interviews to identify the implementation of this GP:
> - How do you identify deviations from the plan? What do you do in case of deviations?
> - How often do you adjust the plan?

GP 2.1.4: Define responsibilities and authorities for performing the process. Responsibilities, commitments, and authorities to perform the process are defined, assigned, and communicated. Responsibilities and authorities to verify process work products are defined and assigned. The needs for process performance experience, knowledge and skills are defined.

Responsibility for activities and work products is agreed with involved staff, also for possible verification measures regarding work products. All responsibilities are documented and communicated.

The terms »responsibilities« and »authorities« are best clarified by way of example: The overall project manager is responsible for completing the project within budget, at the agreed date, with the agreed functionality. He is authorized to:

- Execute business transactions up to a maximum amount of n USD,
- Issue work instructions to project team members,
- Commission external suppliers up to a contract value of n USD.

Explicit assignment and communication of responsibilities is an essential characteristic of Level 2. Without them, the risk of project failure rises in proportion to the number of team members.

> **Notes for Assessors**
>
> It is crucial that responsibilities are documented. Such arrangements are typically included in role descriptions. In practice, documentation of project-specific authorities is rather rare.

> **Notes for Assessors**
>
> The following questions may help in interviews to identify the implementation of this GP:
> - Are the responsibilities for the process assigned and have they been communicated?
> - Who is responsible for the verification of work products?
> - What are the know-how, experience, and skills required by members of staff?

GP 2.1.5: Identify and make available resources to perform the process according to plan. The human and infrastructure resources necessary for performing the process are identified made available, allocated and used. The information necessary to perform the process is identified and made available. The necessary infrastructure and facilities are identified and made available.

In GP 2.1.2, necessary resources for the process performance were planned. Resources refers to »human resources« (i.e., staff) as well as software, hardware, tools and other infrastructure. GP 2.1.5 requires that for the project:

- Suitable staff is identified and made available in sufficient number, for example through the line organization. This includes ensuring that resources are committed.
- Suitable infrastructure (software, computers, miscellaneous hardware, test environments, etc.) and facilities (premises, office furniture, telephones, etc.) are provided.
- If necessary, additional resources are made available through scaling up the plan.
- Information necessary for process control is identified and available (and used).

It hardly makes sense to document staff requirements (i.e., which individual is needed for which period of time) separately for each process. Instead, resource needs are documented and maintained once for the project.

The base practice also covers necessary adjustments regarding resources during changes to the plan (often ramping up). In practice, this often turns out to be rather difficult. Generic practices of capability level 3 (GP 3.2.4 and GP 3.2.5) include similar requirements; there, however, selection and provision of resources must be based on a documented standard process including role descriptions and infrastructure requirements.

Notes for Assessors

Below, problems which commonly arise during the practical implementation of this GP are described and some solutions suggested:

- Assumptions taken at the time of planning often become superceded by reality. It may, for instance, no longer be possible to provide originally planned resources because of miscellaneous deadline changes, or because resources are overbooked. In these cases, realistic adjustments to the plan are necessary, based on up-to-date information.
- Planned use of the same resources in different projects[a] often gives rise to conflicts, and the project manager who »cries out loudest« wins the bid. This can be remedied by a well-functioning resources reservation system in combination with management prioritization in case of conflict.
- Human resources are overbooked (often far exceeding 100 % of the daily work hours over long periods of time). In an assessment, the cross-project workload of project team members must be considered to see if, in the aggregate, they are overbooked beyond capacity.

Consideration must also be given to providing the project with a suitable infrastructure (powerful computers, sufficient software licenses, necessary programs and tools, network access, communication facilities, e.g., for video and telephone conferences), miscellaneous hardware (like test environments and test equipment), and appropriate premises (work places, sufficient meeting rooms, etc.). This can be verified by reviews at staff work places and by on-site visits.

a. Developers, for instance, simultaneously work in several projects allocated to different members of management.

Notes for Assessors

In interviews, the following questions may help to identify the implementation of this GP:

- Which human resources do you actually need? How are they made available to the project? Is it ensured that the project can actually use these resources; i.e., are they reliably allocated?
- Which infrastructure do you need? How is it identified and made available?

*GP 2.1.6: **Manage the interfaces between involved parties**. The individuals and groups involved in the process performance are determined. Responsibilities of the involved parties are assigned. Interfaces between the involved parties[5] are managed. Communication is assured between the involved parties. Communication between the involved parties is effective.*

Systematic management of interfaces includes:

- Regular coordination meetings at planned dates with all required participants.
- Standardized communication solutions in larger projects and distributed teams (e.g., using reporting templates, e-mail distribution lists).

A documented communication plan is recommended for communication planning and control (also called a communication matrix, see figure 3–5).

Communication by mouth is effective, if the usual rules for productive meetings are observed, e.g.,:

- (Only) the right people are invited.
- Invitations are issued in time, stating agenda, start and finishing times.
- Meetings are professionally chaired.
- Decisions are clearly and indisputably recorded.
- The planned meeting duration is kept.
- Compliance with decisions is checked in subsequent meetings.

> **Notes for Assessors**
>
> The following questions may help in interviews to identify if the indicators for this GP have been implemented:
>
> - Who is involved in the process? Who is responsible for what?
> - How are the interfaces managed and how does communication take place?

---

5. »Parties« can be individual persons, or groups.

3.3 The Capability Levels

Communication Plan for Project XYZ

	Information Type	Responsible Person/Author/Role	Area Manager	Department Manager	Team Leader	Product Manager	Designer A	Designer B	Service Department	Production	...	Frequency	Distribution mechanism, email distribution list
Project Management													
	Work Breakdown Structure	Project Manager			X								
	Project Schedule	Project Manager		X	X	X	X	X	X	X			
	Risk List	Project Manager		X	X	X			X	X			
	Communication Plan	Project Manager											
	Project Status Report	Project Manager	X	X	X	X	X	X	X	X		monthly	email distribution list, presentation in meeting
	Final Report/Lessons Learned	Project Manager											
Requirements Management													
	Technical Product Description	Sales											
	System Architecture/System Description	System Architect											
	System Block Diagramm	Designer											
	Legal Framework	Standards Office											
	Requirements Specification	Designer											
	Functional Description	Designer											
Procurement													
	Tender Documents	Responsible Purchaser											
Trials and Test													
	EMC Test												

Figure 3–5   *Communication Plan*

PA 2.2 Work Product Management Attribute

The work product management attribute is a measure of the extent to which the work products produced by the process are appropriately managed. As a result of full achievement of this attribute:

a) *requirements for the work products of the process are defined;*
b) *requirements for documentation and control of the work products are defined;.*
c) *work products are appropriately identified, documented, and controlled;*
d) *work products are reviewed in accordance with planned arrangements and adjusted as necessary to meet requirements.*

The work products of the process are identified. Reviews of the work products are performed and can be verified. Changes are verifiably controlled. Requirements on the respective work products are defined, e.g., regarding content, structures, and quality (as a basis for reviews). Furthermore, requirements on the documentation and control of work products are defined, for instance, regarding traceability, distribution, approval, and configuration control, as needed. Work products to be controlled are identified, put under change and version management, and are readily available. Work products are reviewed according to plan (see GP 2.1.2) against the defined requirements and adjusted if necessary.

Control of work products according to PA 2.2 is supported by quality assurance processes[6], as well as the configuration and change management processes SUP.8 – SUP.10.

> **Notes for Assessors**
>
> In practice, work products deemed necessary by the assessment team are often missing[a], or work products are deficient. First of all, this is negatively reflected in the PA 1.1 rating but also has an effect on the rating of PA 2.2: It often happens that necessary work products are missing, although the the existing work products are well managed according to PA 2.2. Nevertheless, PA 2.2 must be downgraded in these cases because it presupposes the availablity of necessary work products. Otherwise, the organization would gain a wrong impression of its process maturity.

a. By no means all works products required by Automotive SPICE must actually be available. Which work products are appropriate/necessary in a particular context is left to the judgement of the assessor.

GP 2.2.1: *Define the requirements for the work products.* *The requirements for the work products to be produced are defined. Requirements may include defining contents and structure. Quality criteria of the work products are identified. Appropriate review and approval criteria for the work products are defined.*

---

6. SUP.1, SUP.2, SUP.4.

Where work products of the engineering processes are concerned, care should be taken that besides functional requirements (e.g., a navigation system's requirements on the function »route calculation«), nonfunctional requirements are also defined (e.g., performance requirements, compliance with guidelines, accuracy, unambiguousness, consistency, etc.).

Quality criteria can, for instance, be specified by check lists relating to the structure, content, and other quality attributes of a specific work product. Review criteria indicate if and under which conditions a work product needs to be reviewed (and which type of change requires another review), and which criteria are to be checked during the review. It also needs to be determined which criteria[7] must be met by the work product in order to become formally approved (or accepted or released).

Requirements for work products are often distributed across different documents: in each document itself (e.g., reflected in structures and instructions on how to fill in a document template), in the process manuals of the organization, in project requirements specifications, and so on. A list of required work products which includes associated requirements may help the project team to quickly gain an overview of the scope of work.

> **Notes for Assessors**
>
> The following question may help in interviews to identify the implementation of this GP:
>
> - What are your requirements (regarding content, structure, quality, change management, documentation, etc.) for the process's work products?

GP 2.2.2: Define the requirements for documentation and control of the work products. Requirements for the documentation and control of the work products are defined. Such requirements may include requirements for

1. *distribution*
2. *identification of work products and their components*
3. *traceability*

Dependencies between work products are identified and understood. Requirements for the approval of work products to be controlled are defined.

Relevant rquirements are, for instance:

- Who receives which process work product for what purpose (information, review, approval, etc.)? This may, for instance, be specified in a communication plan; see also figure 3–5
- A unique identification method for work products
- Which process work products require agreement/approval, and by whom?

---

7. One criterion could be a successful review.

- How are process work products related to other work products in terms of consistent traceability (supporting, among other things, change management)

Requirements on how dependencies between work products (traceability) are to be tracked vary from process to process. In the engineering processes, base practices explicitly require traceability of product requirements across the entire development cycle (from requirements elicitation all the way to system test).

In non-technical processes, the documentation of all dependencies is often impractical, because economically speaking the complete identification, description and maintenance of all relationships (typically *n:n* relationships) cannot be justified. Project staff, however, should know the relationships between relevant work products in their work environment (for example »Is it known which documents are to be analyzed and modified if document XY is changed?«). Particular consideration must be given to schedule and resource dependencies. For example, the project plan and the quality plan are interdependent. For instance, quality assurance measures planned in the quality plan must also be reflected in the project plan, necessary resources must be available, and so on. For instance, if a new measure is planned in the quality plan, the project plan must be adjusted accordingly (see also GP 2.2.3).

> **Notes for Assessors**
>
> The following question may help in interviews to identify the implementation of this GP:
> - What are the arrangements regarding documentation and control of work products (identification of work products, distribution circle, approval, traceability)?

GP 2.2.3: Identify, document and control the work products. The work products to be controlled are identified. Change control is established for work products. The work products are documented and controlled in accordance with requirements. Versions of work products are assigned to product configurations as applicable. The work products are made available through appropriate access mechanisms. The revision status of the work products may readily be ascertained.

The process work products to be controlled are identified, managed according to the requirements of GP 2.2.1 and 2.2.2, and put under configuration and change management. This includes:

- Work products contain version information and include a change history. The status of each work product is known.
- Changes follow a defined process (e.g., change requests are collected, analyzed, approved, assigned, tracked, and verified).

3.3 The Capability Levels

- Changes become visible in all affected documents and can be traced back from the documents to the original change request.
- Selected work products can be related to corresponding product configurations (e.g., »Which version of the requirements specification is part of this delivery?«).
- Work products are at the disposal of everyone required to use them and are protected from unauthorized access or change.

Although Automotive SPICE does not prescribe the use of tools, practical experience has shown the necessity to use a CM tool, at least for the work products of the engineering processes (e.g., requirements specification, design documentation, code, test cases).

Moreover, if a CM tool is properly used it already covers some base practices of the configuration management process (SUP.8). Some (non-engineering) work documents are often kept in a file system outside the tool. In that case, a corresponding rule-set (e.g., in the CM plan) must exist (see SUP.8). In general, the following applies: The appropriate implementation of the SUP.8 base practices also supports the achievement of GP 2.2.3.

> **Notes for Assessors**
>
> The following questions may help in interviews to identify the implementation of this GP:
>
> - Are work products under configuration and change management? If Yes, which are they? Can versions be uniquely identified? Are changes traceable? Can work products be clearly related to specific product configurations? Are access rights and distribution of the work products defined?

***GP 2.2.4: Review and adjust work products to meet the defined requirements.** Work products are reviewed against the defined requirements in accordance with planned arrangements. Issues arising from work product reviews are resolved.*

Work products are reviewed against the requirements defined under GP 2.2.1. Reviews are performed according to the plan (see GP 2.1.2). Moreover, the generic practice also comprises the resolution of all problems detected during reviews.

Reviews should be conducted according to the four-eye principle (review of a work product by someone other than the author himself). This can be done either by project team members (so-called peer reviews[8]) or by external staff (e.g., QA), or by both (e.g., content reviewed by project team members, formal aspects reviewed by QA). Review results are documented in review records containing,

---

8. The review is conducted by colleagues; senior staff is not involved.

among other things, all detected defects and all resulting actions and associated responsibilities. The removal of problems and defects must be verified through another review (for further information relating to reviews, see SUP.4).

> **Notes for Assessors**
>
> The following questions may help in interviews to identify the implementation of this GP:
>
> - Are reviews or other quality assurance measures performed on the work products of the process? Have these activities been carried out? Is there evidence for this? How are the results documented? How are detected problems removed?

3.3.4 Level 3 (»Established Process«)

The previously described Managed process is now implemented using a defined process that is capable of achieving its process outcomes[9]

In its elaborations on Process Attribute PA 3.1 in section 6.4, ISO/IEC 15504-3 defines the term »defined process« as follows:

A »defined process« is a process that is tailored from the organization's set of standard processes according to the organization's tailoring guidelines ... A project's[10] *defined process provides a basis for planning, performing, and improving the project's tasks and activities.*

While, as long as the requirements of Automotive SPICE are met, projects at Level 2 can utilize very different project-specific processes, processes at Level 3 are tailored from the organization's set of standard processes.

The concept of an organizational standard does not imply that a large organization works according to one single standard process. In a globally operating organization with many thousands of employees, it does not make any sense to require one unified standard process for the many different business areas and development approaches. A standard process should be applicable to an effective organizational unit (e.g., standard process within one customer segment or development area using the same development technologies).

The following assets and documents are used with the standard processes: Documented role descriptions including competency requirements, infrastructure requirements, and metrics to monitor the processes. This does not mean that all projects have to work in exactly the same way. Projects vary in size, development technologies, etc.; factors which are accounted for in the following way:

---

9. This refers to the process outcomes defined in the process dimension for each of the processes.
10. Of course, the same applies to non-project-specific processes.

- Generation of standard process variants: There are different standard processes for different project types. All processes are derived from one common standard. The generation of standard process variants is not explicitly mentioned in Automotive SPICE. However, it is recommended for organizations where due to the diversity of the project types and development methodologies one single standard process is not enough.
- Tailoring: A selected standard process is adapted to meet the specific needs of the project. The possible scope is defined by tailoring guidelines. As a result, there is a »defined process« describing the project's specific process approach.[11]

Figure 3–6 provides an overview of the process attributes and generic practices of Level 3.

Process Attributes		Generic Practices	
PA.3.1	Process definition	GP 3.1.1	Define the standard process that will support the deployment of the defined process.
		GP 3.1.2	Determine the sequence and interaction between processes so that they work as an integrated system of processes.
		GP 3.1.3	Identify the roles and competencies for performing the standard process.
		GP 3.1.4	Identify the required infrastructure and work environment for performing the standard process.
		GP 3.1.5	Determine suitable methods to monitor the effectiveness and suitability of the standard process.
PA.3.2	Process deployment	GP 3.2.1	Deploy a defined process that satisfies the context-specific requirements of the use of the standard process.
		GP 3.2.2	Assign and communicate roles, responsibilities, and authorities for performing the defined process.
		GP 3.2.3	Ensure necessary competencies for performing the defined process.
		GP 3.2.4	Provide resources and information to support the performance of the defined process.
		GP 3.2.5	Provide adequate process infrastructure to support the performance of the defined process.
		GP 3.2.6	Collect and analyze data about performance of the process to demonstrate its suitability and effectiveness.

Figure 3–6   Level 3 process attributes and generic practices

PA 3.1 Process Definition Attribute

The process definition attribute is a measure of the extent to which a standard process is maintained to support the deployment of the defined process. As a result of full achievement of this attribute:

---

11. The transition between the generation of variants and tailoring is blurred; the term »tailoring« is often taken to mean both.

a) *a standard process, including appropriate tailoring guidelines, is defined that describes the fundamental elements that must be incorporated into a defined process;*
b) *the sequence and interaction of the standard process with other processes are determined;*
c) *required competencies and roles for performing a process are identified as part of the standard process;*
d) *required infrastructure and work environment for performing a process are identified as part of the standard process;*
e) *suitable methods for monitoring the effectiveness and suitability of the process are determined.*

NOTE: *A standard process may be used as-is when deploying a defined process, in which case tailoring guidelines would not be necessary.*

Notes for Assessors

The case mentioned in the notes frequently occurs in tendering processes or reporting systems (from projects to management), also in cases where identical processes are essential. It also occurs in organizations aspiring to Level 3 which took short-cuts on tailoring guidelines, or which have not got around yet to creating them. It should therefore be examined very carefully to ensure that the way the process is performed actually complies with the standard process. After all, a well-functioning tailoring process is an essential component of Level 3. Should this not be the case, the process attributes will be downgraded.

GP 3.1.1: Define the standard process that will support the deployment of the defined process. A standard process is developed that includes the fundamental process elements. The standard process identifies the deployment needs and deployment context. Guidance and/or procedures are provided to support implementation of the process as needed. Appropriate tailoring guideline(s) are available as needed.

GP 3.1.1 requires the definition of a standard process, which is valid for a particular part of the development organization. Based on this standard process, projects derive their own project-specific process (the so-called »defined process«).

At Level 3, Automotive SPICE requires a process description which consists of so-called fundamental process elements. This means that the process description includes several different structural elements, such as:

- Process activities with dependencies and interfaces
- Input and output work products
- Support tools and other means
- Information on roles and the way they are involved in the activities (e.g., responsible, contributing, requiring approval (see GP 3.1.3))

The EITVOX[12] model [Radice et al. 1988] is an acknowledged description and modeling method (see figure 3–7).

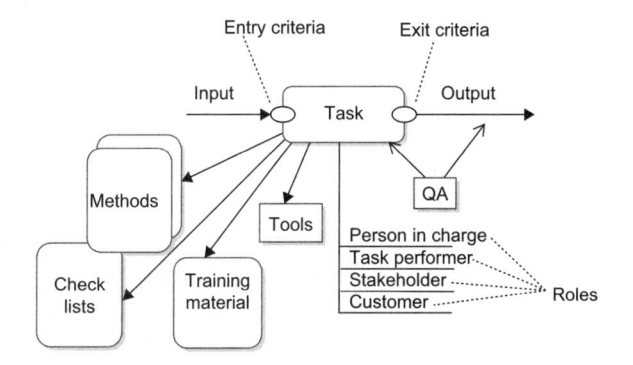

Figure 3–7 Process architecture according to the EITVOX model

A defined standard process requires a sufficiently detailed and, if possible, user-friendly process description accessible to all users. Many organizations use HTML-based presentations, for instance, on their proprietary intranet systems. A well-structured intranet presentation offering an attractive mix of graphical and textual design elements is likely to yield much greater user acceptance compared to conventional textual notations. Hyperlinks also allow faster access to additional information (for instance, to other processes, role descriptions) and documents (e.g., document templates).

Moreover, a process description contains information on possible process deployment areas (application context), i.e., information on the organizational units, project types and/or applications to which the process can be effectively applied. It also states the specific requirements applicable in these cases, for instance, regarding the provision, qualification, and training of staff, or the availability of particular software tools.

The implementation of the process in the project or organization (e.g., in the process groups PIM, REU) is supported by guidelines and procedures. These can be experience reports, examples of good practices, interpretational support, training materials, or checklists.

Tailoring guidelines describe ways in which the standard process can be tailored into a defined process and how tailoring is performed (if, for instance, the tailoring results must be approved).

Process Tailoring

At project start, each process should be tailored from the standard process. A decision must be made which activities are to be carried out and which work products need to be created or adjusted; also, if they may be left out.

---

12. EITVOX stands for Entry criteria, Inputs, Tasks, Verification, Outputs, and eXit criteria.

Project team members need help to derive the defined process in the project[13] from the standard process. Tailoring guidelines must therefore contain clear and unambiguous instructions. Contingent upon project context (for instance, project size, project type, technology) or organizational context, these guidelines should describe what may or may not be changed or omitted. Tailoring not only addresses activities, but also work products. Different project types may, for instance, use different document templates for the same work product.[14]

Tailoring is typically based on categories created for the process's different deployment areas (e.g., project types). It should also be possible to understand the reasons why, for instance, a project was placed in a particular category. For each category, the required activities and necessary work products are specified. If decisions need to be made, the allowed scope is also stated. Such decisions must be traceably documented.

Tailoring tables (figure 3–8 provides an example) are an effective means for tailoring. They tell the project manager which activities need to be performed in his project and which work products are relevant. The example distinguishes between 3 project types (A, B, and C). Depending on the project type, different activities and work products are required. A type C project, for instance, is required to provide a fully filled-in template for the system requirements specification, whereas a type A project may make do with a simplified template.

Process/Activity	Project type A	Project type B	Project type C	Document templates to be used per project type A	Document templates to be used per project type B	Document templates to be used per project type C	PL	Line managmt.	QS	System architect	Requirements engineer	SW developer
Process system requirements analysis												
1.1 Planning of the requirements management/engineering (RM/RE) activities	X	X	X	Project plan	Project plan	RM/RE-plan	A	I	I	I	R	I
1.2 Identify stakeholder involvement		X	X	–	Stakeholder analysis	Stakeholder analysis	R	I			I	
1.3 Creation of the system requirements specification (SRS)	X	X	X	SRS light template	SRS template sections 1, 2, 3, 6, 8	SRS template complete	A		I		R	I

Legend: R: Responsible: drivers of the activity (exactly one R per line)
A: Approval: aacceptance of the results of the activity
I: Informed: will be informed

Figure 3–8   Tailoring table

---

13. Tailoring can be performed in projects. In case of organizational processes, it can also be performed outside projects.
14. For example, project plan templates with different degrees of complexity for different project sizes.

3.3 The Capability Levels

> **Notes for Assessors**
>
> Automotive SPICE processes do not need to map 1:1 to standard processes of the organization. The organization should name and structure its processes as it deems fit and effective for its own business activities. It is the assessors' task to come to an understanding of the mapping through interviews and documentation reviews.
>
> However, it is useful for an organization to create and maintain such a mapping (for instance, in the form of a table in which the organization's processes are related to the Automotive SPICE processes). This does not only render support in assessments but also facilitates the tracking of the correct and complete implementation of the requirements.

> **Notes for Assessors**
>
> The following questions may help in interviews to identify the implementation of this GP:
>
> - Does a documented standard process exist? Are you familiar with it?
> - How can you or other members of the project team access the current version?
> - To which deployment areas does this standard process apply?
> - Are there tailoring guidelines?
> - How do you tailor? Do you receive any support?
> - Do the results need to be approved? If so, by whom?

GP 3.1.2: Determine the sequence and interaction between processes so that they work as an integrated system of processes. The standard process's sequence and interaction with other processes are determined. Deployment of the standard process as a defined process maintains integrity of processes.

What is important here is that the interfaces of the standard process to other processes, and the resulting sequences and interactions, are known and documented. Processes have a reciprocal effect upon each other; they partly depend on one another and possess mutual interfaces, for instance, via

- Work products which are developed or used by several processes
- Process activities of several, tightly inter-connected processes
- Roles carrying out activities in several processes.

The standard process definition of a project management process may, for instance, specify which planning documents of other processes (e.g., quality plan) are to be reviewed and adapted, should the project plan be changed. Or, to give another example, it may prescribe that after defect detection in system test, defect recording and removal must be performed within the incident management process.

The tailoring of the standard process to the defined process must preserve the integrity[15] of the process as a whole, i.e., the interfaces and interconnected processes mentioned above must remain intact even after tailoring.

> **Notes for Assessors**
>
> The following questions may help in interviews to determine the implementation of this GP:
> - How does the standard process relate to other processes?
> - Are there any interfaces between the processes?
> - Are there any interacting processes/procedures?

GP 3.1.3: Identify the roles and competencies for performing the standard process[16]. *Process performance roles are identified. Competencies for performing the process are identified.*

A standard process description also contains information on the roles (e.g., project manager, QA representative, developer) involved in the different activities of the process. This is typically done in the form of a responsibility matrix assigning different types of responsibility to individual roles.[17]

In addition, role descriptions must be prepared, specifying:
- The way in which a role is involved in activities,
- The scope of responsibilities and authorities
- The competencies required to fulfill the role.

Competence requirements can be divided into technical or domain competence, methodological competence, and social competence. Figure 3–9 provides an example of a role description for a software project manager.

---

15. i.e., soundness, completeness.
16. Strictly speaking, the defined process, not the standard process, is performed.
17. An example of a responsibility matrix is included in the tailoring table in figure 3–8.

3.3 The Capability Levels

	Software (Sub) Project Manager
Role Description	The software (sub) project manager is responsible for the planning and control of all software development activities, including project planning and tracking, as well as cost control. He/She manages the software team. A software project manager is assigned for each project that includes software development.
Responsibilities	■ Selects the team members in agreement with the (overall) project manager ■ Ensures that software related project goals are achieved through appropriate planning and tracking of project activities ■ Ensures that all necessary resources are available for the performance of the software activities ■ Releases software deliveries
Activities	1. Controls the software (sub) project to achieve the project goals with regard to software activities defined in the project definition. This includes: 　■ Planning and defining the project scope relating to software, documentation in the software development plan. This includes: Definition of the work packages in cooperation with involved team members, budget planning, scheduling, planning of QA activities including reviews, test planning, planning of the technical infrastructure ... 　■ Status control and plan-vs-actual comparisons regarding schedules, costs, as well as performance of the planned activities according to the software development plan 　■ Creation and distribution of the project status report regarding software activities 　■ Preparation of milestone reviews and steering committee meetings in agreement with the overall project manager ... 　■ ... 2. Represents the software parts in the overall project, including: 　■ Participation in project management meetings 　■ Participation in the Change Control Board 　■ Selection of second-tier suppliers in cooperation with the (overall) project manager 　■ ...
Competencies	Technical: Very good knowledge of project management and software engineering, basic knowledge of managerial economics, good knowledge of incremental software development, good working knowledge of Microsoft Project Methodological: Strong presentation skillset and experienced facilitator Social: Effective communication skills, conflict management

Figure 3–9　Example of a role description for a software project manager

> **Notes for Assessors**
>
> At Level 3, documented role descriptions stating the responsibilities and necessary competencies are explicitly required and must be evidenced.

> **Notes for Assessors**
>
> The following questions may help in interviews to identify the implementation of this GP:
> - Which roles are involved in the standard process?
> - Which competencies are needed for this role? Are they documented?

GP 3.1.4: Identify the required infrastructure and work environment for performing the standard process. *Process infrastructure components are identified (facilities, tools, networks, methods, etc). Work environment requirements are identified.*

Although not explicitly addressed at Level 1, the availability of a suitable infrastructure and work environment forms the basis on which processes are performed. Otherwise, the purpose of the processes is not achieved. This implies that at Level 1, a suitable infrastructure and work environment is already required. At Level 2, infrastructure and work environment must be explicitly planned for each process as part of PA 2.1 (particularly: GP 2.1.5). At Level 3, infrastructure requirements and requirements on the work environment are already included in the standard process (for instance, the standard process could require the existence of a licence for a requirements management tool). GP 2.1.5 lists further examples for possible infrastructure components.

> **Notes for Assessors**
>
> The following questions may help in interviews to identify the implementation of this GP:
> - Which infrastructure is needed for the standard process?
> - Which work environment is needed? Is it documented?

GP 3.1.5: Determine suitable methods to monitor the effectiveness and suitability of the standard process. *Methods for monitoring the effectiveness and suitability of the process are determined. Appropriate criteria and data needed to monitor the effectiveness and suitability of the process are defined. The need to establish the characteristics of the process is considered. The need to conduct internal audit and management review is established. Process changes are implemented to maintain the standard process.*

The effectiveness and suitability of the standard process can only be judged based on the practical application of the project's or organization's defined process. Suitable monitoring methods are:

- Regular measurements of key figures for applied processes. Collecting key figures for each process does not always make sense. Key figures are process-specific parameters supposed to contribute towards a better understanding of applied processes (for example, the number of requirements changes per time unit in a requirements management process).
- Management reviews: Management reviews can review the contents and implementation of applied processes to see whether they are appropriate and effective. This can be done in many ways. A practical form, for instance, would be the performance of regular (e.g., every 2–3 months) »process steering committees« meetings. Issues discussed in these meetings are the progress of process improvement activities, as well as technical and formal issues (up to a certain degree of detail). Barriers impeding the implementation of processes are removed. Review participants may be managers from the different management levels in development, process owners[18], QA representatives, as well as the project manager of current process improvement projects.
- Internal audits (Review regarding process compliance, e.g., internal process audits conducted by QA)
- Feedback from project team members, for instance, subsequent to post mortem evaluations at project completion[19]

GP 3.1.5 requires the definition of such methods in conjunction with the standard processes. GP 3.2.6 describes the way in which these methods are applied in a process. GP 3.2.6 thus uses the methods described in GP 3.1.5.

The evaluation as to whether a standard process is effective and suitable must be based on criteria relating to data obtained from the assessment methods that we just mentioned. It may, for instance, be considered an alarm sign if a particular process shows a significantly lower process compliance (identified by internal audits) compared to other processes. Or, to give another example, the unit test process may require revision if at least one of the following criteria applies:

- The practicability of the method was questioned by developers during the evaluation of experiences in more than one project.
- In process audits, the percentage of process non-conformance is at least 20 % higher than in other processes.
- More than 15 % of defects detected during integration test are defects that should have been detected during unit testing.

Measured key figures can also be used for the evaluation of process effectiveness and suitability. For instance, to assess the effectiveness and suitability of a review

---

18. Process owners are responsible for the definition and maintenance of processes.
19. It is good practice to offer staff members the opportunity to give feedback regarding processes (problems, insights, suggestions for improvement) and to provide defined reporting mechanisms (e.g., via intranet forms to process representatives).

method, the average number of problems detected in reviews can be compared with experience documented in the literature, or with figures of other organizations.

Based on these findings, the standard process is continuously improved. Improvement of the standard process is an important attribute of Level 3, since processes age very quickly and require continuous maintenance effort. If this is neglected, acceptance and process compliance will deteriorate rapidly.

> **Notes for Assessors**
>
> Evidence should be examined to see if, based on insights gained from the methods mentioned, maintenance of the standard process has actually taken place.

As regards measuring data, it is good practice (but not required in GP 3.1.5) to collect, prepare, analyze, and store data continuously over time. A »historical database« is a valuable asset, e.g., to compare current measurement data with historical data, or to prove conclusively that process improvements have been made.

In GP 3.1.5, Automotive SPICE does not explicitly require continuous process improvement; it does, however, require process maintenance. GP 3.2.6 also only identifies opportunities for process improvement (refer also to the elaborations in GP 3.2.6).

> **Notes for Assessors**
>
> The following questions may help in interviews to identify the implementation of this GP:
> - Which methods are used to evaluate the effectiveness of the standard process?
> - Is feedback being collected regarding the improvement of the standard process?
> - Does management perform reviews of the processes?
> - How is the process maintained?

PA 3.2 Process Deployment attribute

The process deployment attribute is a measure of the extent to which the standard process is effectively deployed as a defined process to achieve its process outcomes. As a result of full achievement of this attribute:

a) *a defined process is deployed based upon an appropriately selected and/or tailored standard process;*

b) *required roles, responsibilities and authorities for performing the defined process are assigned and communicated;*

c) *personnel performing the defined process are competent on the basis of appropriate education, training, and experience;*

d) required resources and information necessary for performing the defined process are made available, allocated and used;
e) required infrastructure and work environment for performing the defined process are made available, managed and maintained;
f) appropriate data are collected and analysed as a basis for understanding the behavior of, and to demonstrate the suitability and effectiveness of the process, and to evaluate where continuous improvement of the process can be made.

Once the practices of PA 3.1 are implemented, standard processes and information regarding roles, competencies, infrastructure requirements and monitoring methods are available. Using the practices of PA 3.2, the processes are applied in concrete contexts. This means:

- The so-called »defined process« is tailored from the standard process.
- Roles, responsibilities and authorities are assigned and communicated.
- Deployment of competent staff is based on explicit competence requirements.
- Resources, information, infrastructure, or work environment are provided, based on the requirements. Maintenance of infrastructure and work environment is ensured.
- Measurements are collected during the performance of the defined process so that the process can be better understood and improved.

Notes for Assessors

Traceable evidence must be available on the tailoring of the process, and the resulting defined process must be appropriately documented. In addition, it needs to be examined whether the project's defined process is being performed in a verifiable and consistent manner.

GP 3.2.1: Deploy a defined process that satisfies the context specific requirements of the use of the standard process. The defined process is appropriately selected and/or tailored from the standard process.[20] *Conformance of defined process with standard process requirements is verified.*

Based on the standard process defined under GP 3.1.1, a »defined process« is derived, documented, and applied[21]. If there are several possible standard processes (if, for instance, process variants exist), the process must be derived from the one that is most suitable. This is done using documented tailoring guidelines. Tailoring decisions and necessary adjustments must be justified and documented.

---

20. By means of tailoring.
21. The process may be applied in projects or in organizational processes outside of projects.

Proof must be provided that the requirements of the standard process were not violated and that it was tailored in compliance with the tailoring guidelines. This can be established by review or approval through the QA team.

> **Notes for Assessors**
>
> The following questions may help in interviews to identify the implementation of this GP:
>
> - How was the process you are using derived from the standard process? How were the results documented? Has there been a check to determine conformity with the standard process?

GP 3.2.2: Assign and communicate roles, responsibilities and authorities for performing the defined process. *The roles for performing the defined process are assigned and communicated. The responsibilities and authorities for performing the defined process are assigned and communicated.*

The assignment of roles to individuals defined in GP 3.1.3 is documented and communicated. If not already defined by roles, the responsibilities and authorities are also documented and communicated.

> **Notes for Assessors**
>
> The following questions may help in interviews to identify the implementation of this GP:
>
> - Which roles, responsibilities and authorities were assigned to members of staff? How was this documented and communicated?

GP 3.2.3: Ensure necessary competencies for performing the defined process. *Appropriate competencies for assigned personnel are identified. Suitable training is available for those deploying the defined process.*

The competencies of staff designated for a role are verified against the competence requirements defined in GP 3.1.3. It is of help if the organization maintains relevant records (qualification records, data bases, etc.) on the competencies of existing staff. The same applies to special, project-related competencies (e.g., regarding domain know-how, knowledge of methods, tool knowledge).

If there are any discrepancies, the resulting training requirements are systematically identified, training is planned in due time, and actual participation of staff in relevant training programs is ensured. Although not required by Automotive SPICE, it is also good practice to verify the quality of the training, for instance, by using evaluation sheets to be completed during employee surveys after training has finished. The evaluation sheets are analyzed and, if necessary, improvement measures are initiated.

> **Notes for Assessors**
>
> Check if a comparison between required and existing competencies was made and if, in consequence, training was planned and actually attended.

> **Notes for Assessors**
>
> The following questions may help in interviews to identify the implementation of this GP:
> - Which competencies are required in this process? Was a comparison made between required and available competence? Were training programs derived? Did training take place?

GP 3.2.4: Provide resources and information to support the performance of the defined process.[22] *Required human resources are made available, allocated and used. Required information to perform the process is made available, allocated and used.*

GP 2.1.5 at Level 2 already required that resources be identified and made available. This requirement is exceeded by the generic practices GP 3.2.4 and GP 3.2.5, because at Level 3 there are standard processes (GP 3.1.1) with defined roles (GP 3.1.3). Resources are assigned to these roles (GP 3.2.2), they are adequately qualified (GP 3.2.3) and systematically and reliably allocated and made available (GP 3.2.4). There are infrastructure requirements for the necessary infrastructure (GP 3.1.4), and the infrastructure is systematically and reliably made available and maintained according to the requirements (GP 3.2.5).

In GP 3.2.4, resources are selected that are principally suitable on the grounds of competency, or which can be trained in terms of GP 3.2.3. Here it becomes clear that GP 3.2.2, GP 3.2.3, and GP 3.2.4 can only be implemented in conjunction with each other (and not necessarily in the sequence suggested in Automotive SPICE).

> **Notes for Assessors**
>
> Verify proof that members of staff are actually available according to plan. This can be established, for instance, by checking the hours entered in the time recording system or through interviews with project team members.

The kind of information necessary for the performance of a defined process depends on the particular process. Relevant information must be provided

---

22. The use of the verb »support« could give the wrong impression that Automotive SPICE refers to »support personnel«. However, it refers to personnel for the performance of the process (see Process Outcomes (d) in PA 3.2).

systematically, for instance, in the form of a well structured project directory or as data stored in an intranet repository. The information should be up-to-date and easily accessible by all involved. Distributed development poses enhanced requirements. Examples of the type of information required are:

- Examples and experience from projects that are already completed
- Estimation models for effort estimates, based on historical data
- A data base with historical measurement data related to different project and organizational processes
- Process-related training material

> **Notes for Assessors**
>
> The following questions may help in interviews to identify the implementation of this GP:
>
> - How were members of staff allocated and made available? Is staff allocation reliable?
> - Which support information is available to assigned resources (estimation models, historical data, training documents etc.)?

GP 3.2.5: *Provide adequate process infrastructure to support the performance of the defined process.* [23] *Required infrastructure and work environment is available. Organizational support to effectively manage and maintain the infrastructure and work environment is available. Infrastructure and work environment is used and maintained.*

The standard infrastructure defined in GP 3.1.4 (as well as additional infrastructures not mentioned there) must be made available in the defined process (e.g., in the project). If organizational support is needed for the use or maintenance of the infrastructure (e.g., technical software and hardware support), it will be provided. The defined infrastructure is actually used in the projects.

The delimitation to the generic practice GP 2.1.5 is described in GP 3.2.4.

> **Notes for Assessors**
>
> The following questions may help in interviews to identify the implementation of this GP:
>
> - Which infrastructure is required? Is it available?
> - How was suitability determined and on what grounds?
> - Is it documented? Is infrastructure support available?

---

23. Regarding the term »support«, refer to the previous footnote.

GP 3.2.6: Collect and analyse data about performance of the process to demonstrate its suitability and effectiveness. Data required to understand the behavior, suitability and effectiveness of the defined process are identified. Data are collected and analysed to understand the behavior, suitability and effectiveness of the defined process. Results of the analysis are used to identify where continual improvement of the standard and/or defined process can be made.

GP 3.1.5 requires the definition of methods in conjunction with the standard processes. Their use in the defined process is described in GP 3.2.6. According to the methods defined in GP 3.1.5, regular measurements, reviews and audits are performed in the defined process. The results are analyzed in order to understand the behavior of the process and to be able to assess its suitablity and effectiveness. Measurement data is regularly collected and analyzed. Based on the analysis, process improvement potentials are identified in the defined and standard processes. In this connection, trend analyses are good practice, i.e. the prediction of a future trend based on past and current data.

GP 3.2.6 does not require that any improvements are made. It only requires that improvement potentials are identified. At Level 3, Automotive SPICE does not explicitly require continuous process improvement. This is required at Level 5 and in the PIM.3 process. PA 3.1 and PA 3.2 are only designed to identify improvement potentials, and therefore render valuable preliminary work for Level 5 and/or the PIM.3 process. Of course, in practice, improvement measures should be carried out independent of the capability level in both the defined process and the standard process. This can be accomplished even with a partial implementation of the PIM.3 process.

> **Notes for Assessors**
>
> The following questions may help in interviews to identify the implementation of this GP:
> - Which data is collected to allow a better understanding of process behavior and to evaluate its suitablity and effectiveness?
> - As a result, have improvement potentials been identified?

3.3.5 Level 4 (»Predictable Process«)

The previously described Established process now operates within defined limits to achieve its process outcomes.

At Level 4, detailed measurements are performed and analyzed while performing the defined process. This results in a quantitative understanding of process performance and an improved prediction accuracy. Using statistical methods, the defined process is controlled within upper and lower control limits, in order to achieve quantitative process objectives and to support the achievement of the business goals. In case of deviations from the control limits, the causes are identified and corrective measures are performed.

At Level 4, the measurement of processes requires an effective measurement system to collect indices for the performance of the process and the quality of the work products, and to support analyses. The applied analysis and control techniques depend on the process and organizational unit. Statistical process control is not suitable for every process and does not always make sense. Statistical process control, for instance, makes more sense in engineering than in organizational processes.

Figure 3–10 provides an overview of the Level 4 process attributes and generic practices.

Process attributes		Generic practices	
PA.4.1	Process measurement	GP 4.1.1	Identify process information needs
		GP 4.1.2	Derive process measurement objectives
		GP 4.1.3	Establish quantitative objectives
		GP 4.1.4	Identify product and process measures
		GP 4.1.5	Collect product and process measurement results
		GP 4.1.6	Use the results of the defined measurement
PA.4.2	Process control	GP 4.2.1	Determine analysis and control techniques
		GP 4.2.2	Define parameters suitable to control the process performance
		GP 4.2.3	Analyze process and product measurement results
		GP 4.2.4	Identify and implement corrective actions
		GP 4.2.5	Re-establish control limits

Figure 3–10 Level 4 process attributes and generic practices

3.3.6 Level 5 (»Optimizing Process«)

The previously described predictable process is continuously improved to meet relevant current and projected business goals.

At Level 5, there exists a long-term strategy for the successful implementation of process improvements. Based on the organization's business objectives, areas of improvement and trends are set, and a long-term development strategy is defined.

Suitable improvement opportunities (e.g., from innovative approaches and techniques) are systematically identified, analyzed and piloted. During piloting, the benefit of the improvement is quantitatively identified through measurements and compared with historical data. Improvements that have proved to be suc-

3.3 The Capability Levels

cessful are systematically integrated and implemented in the organization's process landscape.

Level 5 is essentially based on the quantitative understanding of process performance (attribute of Level 4). In addition to the attributes of Level 4, it requires the following:

- Focus on proactive, continuous process improvements that aim at achieving the targets set by present and future business goals. Corresponding activities and resources are planned.
- Process changes are identified, planned, and introduced in an orderly fashion. Care is taken that the performance of the process (the operational business) is interfered with as little as possible. For instance, criticality of the project, planned increases in effectiveness, and the generation of new business areas need to be considered.
- The effectiveness of process changes is quantitatively evidenced by comparison with the previous situation.

Figure 3–11 provides an overview of the process attributes and generic practices of Level 5.

Process Attributes		Generic Practices	
PA.5.1	Process innovation	GP 5.1.1	Define the process improvement objectives
		GP 5.1.2	Analyze measurement data
		GP 5.1.3	Identify improvement opportunities of the process based on innovation and best practices
		GP 5.1.4	Derive improvement opportunities of the process from new technologies and process concepts
		GP 5.1.5	Define an implementation strategy based on long-term improvement vision and objectives
PA.5.2	Process optimization	GP 5.2.1	Assess the impact of each proposed change
		GP 5.2.2	Manage the implementation of agreed changes
		GP 5.2.3	Evaluate the effectiveness of process change

Figure 3–11 Level 5 process attributes and generic practices

4 CMMI – Differences and Similarities

This chapter deals with the differences and similarities between CMMI and SPICE (or Automotive SPICE) from three different angles: structures (section 4.2), content (section 4.3), and assessment or appraisal methods (section 4.4). Section 4.1 provides a brief introduction into the purpose of this model comparison and some of the problems that this entails.

4.1 Introduction

The following elaborations do not intend to balance the advantages and disadvantages of one model against the other, nor will they provide desision support and arguments for or against the selection of one particular model.[1] We rather assume that organizations and their cooperation partner already apply certain, to some extent different, models and that this gives rise to a number of uncertainties and questions. Frequently asked questions are:

1. We internally apply CMMI but assess our own suppliers according to Automotive SPICE. Is this possible, and does it make sense?
2. We assess our suppliers based on Automotive SPICE. A particular supplier internally applies CMMI and provides us with corresponding findings from appraisals. Can/should we recognize these findings and waive supplier assessments altogether or in part?
3. We work closely with a cooperation partner who applies a different model than we do. Does this cause problems, and what do we have to look out for?

On question 1: Yes, of course it is possible and it definitely makes sense. Automotive SPICE for supplier evaluation is standard anyhow. Only the processes of the supplier, not your own processes, are assessed. Therefore, the application of different models is not an issue.

---

1. Some rudimental criteria are listed in chapter 1.

On question 2: A very short answer would be: Yes and No! A more precise answer requires differentiated consideration (see also section 4.3): First and foremost, it is important to know if the supplier has based his process improvement initiative and appraisals on a process scope that can be compared to the HIS scope. This is, for instance, not the case if the supplier has maturity Level 2 (see section 4.2 for details). Even if the supplier has maturity Level 3, there is the problem that two of the Automotive SPICE-processes (change request management and problem resolution management) required by HIS are only partly addressed in CMMI. There are, in addition, several differences of detail between the two models. In many points, Automotive SPICE makes much more concrete demands on requirements which go far beyond those of CMMI (see also section 4.3). Their fulfillment is naturally not an issue in a CMMI appraisal. However, viewed with a little more distance, one may say that despite their differences both models represent effective tools for improving process and product quality. In the case of CMMI there are numerous well-founded accounts of its effectiveness over many years. Besides, CMMI also makes demands on requirements which do not occur in Automotive SPICE. In order to save effort, supplier assessments could be limited to a few sample checks.

One of the problems in this situation is that assessors are confronted with concepts, terminology, structures, and documents at the supplier's that they do not know. As a result, they have difficulties evaluating and classifying them. It is therefore a suitable measure to have one's own assessors sufficiently trained in CMMI.[2] The elaborations in the remainder of this chapter provide at least a rough overview on these issues.

On question 3: The answer depends on the kind and intensity of the cooperation. In a typical customer-supplier relationship there is no full integration of development processes. This means that both partners work based on their own processes and that they cooperate via defined interfaces (such as document formats, strictly defined information flows, and other conventions). Moreover, on the customer's side, supplier management is performed. This is required by both models and leaves the methods by which the supplier works more or less open. This is different in the case of closely interlinked development partnerships in which the working methods of both parties need to be much more closely adapted to each other. These adaptations can then be represented in the form of documented tailoring of the respective standard processes. Care should be taken not to violate the requirements of the reference models.

---

2. Experience has shown that a two-day training course that also looks at the similarities and differences of the two models is sufficient.

4.2 Comparison of Structures

There is a whole range of related CMMI models for different application areas. They constitute the so-called CMMI constellations:

- CMMI for Development (CMMI-DEV): This is the oldest and most widely distributed constellation (the latest version was published in 2006) with focus on development processes. It is structured in a very generic way to make it suitable for very diverse application domains. Explicit examples are provided for software, hardware, and systems. It is easily possible to apply it to other domains, too, for instance, to the development of services, manufacturing processes, and so on.
- CMMI for Acquisition (CMMI-ACQ): This constellation (first publication 2007) is designed for organizations that purchase products or services on a large scale. This constellation would, for instance, suit the situation of the OEMs but does not contain any development processes.
- CMMI for Services (CMMI-SVC): This constellation (first publication planned for 2008) addresses the operation of services of all kinds. Important areas of application are IT services (operation of computer centers), call centers, health services, transportation and traffic, tourism, and so on.

The constellations share a common core of identical process areas and can therefore be used in combination. CMMI-DEV is primarily suited for a comparison with Automotive SPICE, since only there do we find corresponding development processes. CMMI-ACQ also qualifies with regard to the acquisition processes which are very pronounced there and which are only relatively briefly dealt with in CMMI-DEV. In each of the CMMI models, two closely interlinked [3] types of representation are to be distinguished:

- The Continuous Representation is structured in analogy to ISO 15504, i.e., the process areas are divided into categories and the capability levels are identical. Capability levels refer to the respective process area, i.e., each process area can assume an individual capability level. Only this form is suited for comparison with Automotive SPICE.
- The Staged Representation represents a somewhat different philosophy: process areas are combined in defined groups which are supposed to be jointly implemented. Only after successful implementation of a group (and having thus reached a capability level) will the next group be addressed. The groups and sequence are based on long-standing experience. This kind of capability level (here called a »Maturity Level«) does therefore not refer to one individual process area but rather to the collective implementation of a predefined

---

3. Both are represented as integated parts in a cohesive model text.

number of process areas. This is the most widely used form of representation among CMMI users (approx 80%)[4].

Apart from the fact that »Change request management« and »Problem resolution management« do not exist in CMMI as separate process areas, the Automotive SPICE processes that correspond to the HIS scope are addressed at Maturity Level 3. This level, however, contains a whole range of additional process areas.

At first glance, a comparison of SPICE or Automotive SPICE with CMMI (Continuous Representation) does indeed show basic congruities (especially the capability levels), but there are also differences. The SPICE process model, for instance, is more finely structured (48 processes in SPICE and 31 in Automotive SPICE[5] versus 22 process areas in CMMI). Figure 4–1 shows a comparison of the most important terms and concepts. For some concepts there are no counterparts, others can be mapped quite well. The correspondence between Specific Goals and Process Outcomes is relatively weak.

CMMI	SPICE	Automotive SPICE
Process Area	Process	Process
Purpose	Process Purpose	Process Purpose
Specific Goals	Process Outcomes	Process Outcomes
Specific Practices	Base Practices	Base Practices
Subpractices		
Typical Work Products	Input/Output Work Products	Output Work Products
	Work Products Characteristics	Work Products Characteristics
Generic Goals	Process Attributes	Process Attributes
Generic Practices	Generic Practices	Generic Practices
Generic Practice Elaborations		
	Generic Resources	Generic Resources
	Generic Work Products	
Examples		
Amplifications		

Figure 4–1   Comparison of the most important terms and concepts

---

4. As measured by the appraisals that used the Staged or Continuous Representation (see [Phillips 2006]).
5. Both models may also be seen in combination: As mentioned in the Foreword to Automotive SPICE, missing processes may be directly taken over from SPICE.

4.3 Comparison of Contents

Both process models include content not included in the other model. For instance, maintenance and problem resolution management are not addressed in CMMI, and in SPICE/Automotive SPICE various concepts are missing, for instance, »Decision Analysis and Resolution« and »Integrated Product and Process Development«. Moreover, the degree of detail and elaborateness are higher in CMMI. As regards the engineering processes, SPICE (more accurately: ISO/IEC 15504-5) and Automotive SPICE are very much software focused. CMMI, on the other hand, phrases all practices in such a general way that in principle they fit many application domains. However, this makes their concepts relatively abstract. CMMI tries to counteract this with discipline-specific elaborations (for hardware, software, and systems engineering)

SPICE/ Automotive SPICE	Meaning	CMMI	Meaning
Process Attributes	Based on the evaluation of the Process Attributes, the capability levels are calculated.	Goals (Specific Goals, Generic Goals)	Based on the goal ratings, the capability levels are assigned. Goals have highest relevance (status »required«).
Practices (Base practices, generic practices)	The SPICE norm neither requires nor mentions ratings of the practices. However, most SPICE-compliant assessment methods evaluate the practices, too, in order to derive the evaluation of the process attributes from them.	Specific practices, generic practices	Goal ratings are derived from the »characterization« of the practices. Practices are of second-highest relevance (status »expected«).
(no equivalent)		Subpractices (and further model components of informative character)	These model components have the least relevance (status »informative«) and are not evaluated. They do, however, serve as an interpretational aid regarding the practices.

Figure 4–2   Varying relevance of SPICE/Automotive SPICE and CMMI model components in comparison

Wheras SPICE uses a two-tier hierarchy of meaning in its requirements on SPICE compliant assessment methods, consisting of process attributes and practices (see figure 4–2), CMMI uses a three-tier hierarchy in connection with its appraisal method SCAMPI. The vast majority of CMMI text is of an informative character (i.e., lowest hierachy level) and is therefore not concretely evaluated. It has, on the other hand, an impact on the expectations and scope of interpretation. What makes a comparison between CMMI and SPICE more difficult is the fact that many SPICE requirements are actually mentioned in CMMI, but only in the informative part (for instance, in the form of »Subpractices«). Strictly speaking, SPICE requirements are often harder and more precise. In our experience, many of them have been realized in typical CMMI implementations, yet this issue cannot be answered in a pure model comparison but must be examined in the concrete implementation. It is therefore advisable for organizations that work CMMI-compliant to create a corresponding mapping between their processes and Automotive SPICE practices. This way the organization can recognize gaps and prepare for Automotive SPICE assessments[6]. In the remainder of this section we show the most important Automotive SPICE requirements of the HIS scope, which are not *required* in CMMI[7] or in less detail (but which may be mentioned in the informative text). For the meaning of the CMMI abbreviations, refer to figure 4–3.

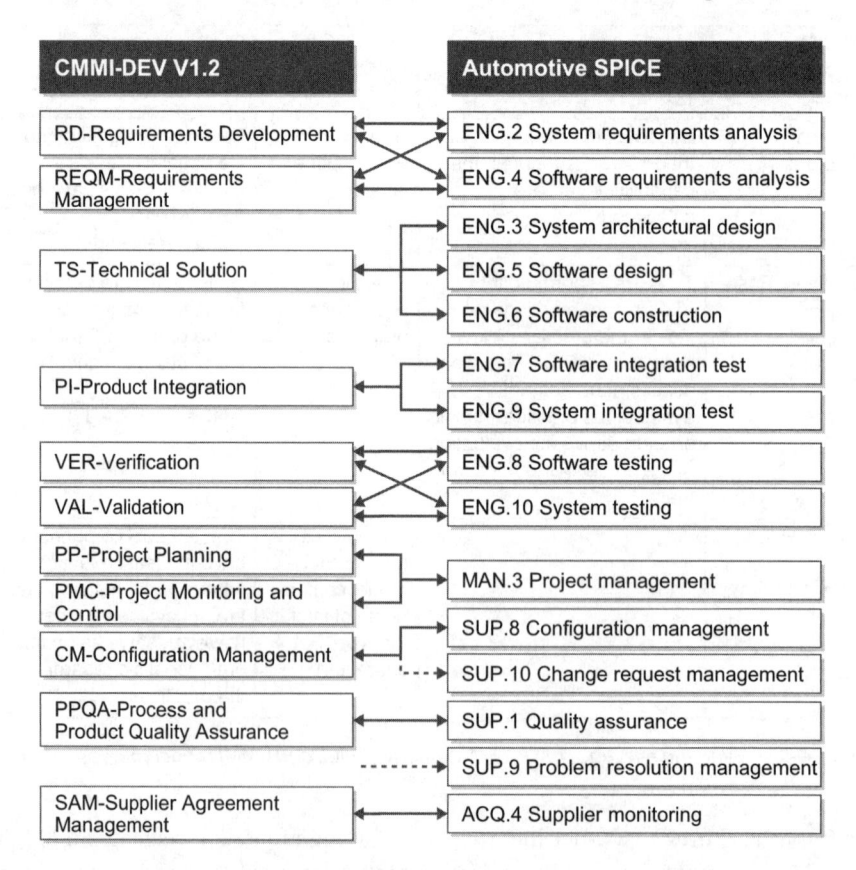

CMMI-DEV V1.2	Automotive SPICE
RD-Requirements Development	ENG.2 System requirements analysis
REQM-Requirements Management	ENG.4 Software requirements analysis
TS-Technical Solution	ENG.3 System architectural design
	ENG.5 Software design
	ENG.6 Software construction
PI-Product Integration	ENG.7 Software integration test
	ENG.9 System integration test
VER-Verification	ENG.8 Software testing
VAL-Validation	ENG.10 System testing
PP-Project Planning	MAN.3 Project management
PMC-Project Monitoring and Control	
CM-Configuration Management	SUP.8 Configuration management
	SUP.10 Change request management
PPQA-Process and Product Quality Assurance	SUP.1 Quality assurance
	SUP.9 Problem resolution management
SAM-Supplier Agreement Management	ACQ.4 Supplier monitoring

Figure 4–3 *Correspondence between CMMI process areas and Automotive SPICE processes (HIS scope only)*

---

6. Another important preparatory step is the performance of an Automotive SPICE trial assessment.
7. There are more differences in detail which due to space constraints cannot be shown here.

RD, REQM compared with ENG.2

- Communication mechanisms for dissemination of requirements (ENG.2, ENG.4) are not required in CMMI.
- Traceability requirements are more explicit in Automotive SPICE.
- Generally: Whereas Automotive SPICE describes three explicit abstraction levels of requirements (customer, system, software), CMMI only uses two (customer, product), whereby »product« can be sub-divided into any number of levels.

SAM compared with ACQ.4

- Automotive SPICE requires more details regarding the cooperation between customer and supplier: common processes and interfaces, exchange of information, review activities, tracking of open points, and correction of deviations.

PPQA compared with SUP.1

- CMMI requires a quality assurance strategy and plan at Level 2, Automotive SPICE at Level 1.
- Explicit Automotive SPICE requirements for a quality organization that is independent from the project organization is not required in CMMI. CMMI only requires objectivity.
- Escalation mechanisms are not required in CMMI.

CM compared with SUP.8, SUP.10

- In CMMI, configuration management strategy is required at Level 2, in Automotive SPICE at Level 1.
- Branch management strategy is not required in CMMI.
- Managing backups, storage, archiving, handling, and delivery are not required in CMMI.
- The SUP.10 requirements (change management) are only partly reflected in the informative text part of a specific practice of the CM process area.

TS compared with ENG.3, ENG.5, ENG.6

- Verification of design and implementation, as required in Automotive SPICE, are not required in CMMI within the TS process area. This is covered by the process areas VER/VAL. The question of which objects are to be verified or validated is left open so that the user can choose for himself.
- The communication mechanisms for the dissemination of the design, required at Level 1 in Automotive SPICE, are more indirectly covered in CMMI by a generic practice at Level 2 (»Stakeholder Involvement«).
- Traceability requirements are more explicit in Automotive SPICE. They are not included in the CMMI TS process area, but are found in REQM.
- A description of the dynamic behavior is not required in CMMI.

- A description of the objectives regarding resource consumption (ENG.5) is not required in CMMI.
- Test criteria (ENG.5) and unit verification strategy (ENG.6) are not included in the CMMI TS process area, but are found in VER/VAL.
- Generally: Whereas Automotive SPICE describes three explicit abstraction levels of design and implementation (system architectural design, software design, software construction), CMMI uses two explicit levels (design, implementation). The distinction of different design levels is described in the informative text.

PI compared with ENG.7, ENG.9

- Automotive SPICE has many more and more detailed requirements regarding the execution and documentation of tests.
- Traceability requirements are more explicit in Automotive SPICE. They are not included in the CMMI PI area, but are found in REQM.

4.4 Comparison of the Assessment/Appraisal Methods

An obvious difference is already apparent in the use of different terms : SPICE/Automotive SPICE talk about »assessment«, in CMMI the official term to be used is »appraisal«[8]. An appraisal is an official evaluation that is registered with the SEI and which can only be performed by an SEI-authorized person. This person must be registered with an SEI partner who in turn must fulfill certain requirements, must be licenced for this kind of service, and pay corresponding licensing fees. This kind of appraisal requires that certain documents must be sent to the SEI [SEI]. They are subject to quality assurance and the results are anonymized and logged in the SEI database in order to be able to generate statistical evaluations. The results are kept strictly confidential and are therefore not available to the public nor to other SEI-authorized persons. The organization concerned, however, does have the opportunity to publish its results [Published Appraisal Results]. It may be interesting to have a look at the list of SEI-authorized Lead Appraisers and the licensed SEI partner list [SEI Partner Directory].

For each SCAMPI appraisal at least the following documents must be created (and reported to the SEI): planning documents, the final findings presentation and the »Appraisal Disclosure Statement« (ADS). The ADS contains standardized information and specifies, among other things, the exact subjects that were appraised, the team organization, who the SEI-authorized person was, what the findings were, and for which part of the organization these findings are valid. It

---

8. In the context of CMMI, if one speaks of »assessment« , it usually refers to a examination that has not been performed according to SEI rules but has, for instance, been performed using an unofficial method or by a person without SEI authorization.

4.4 Comparison of the Assessment/Appraisal Methods

can therefore be very interesting for a (potential) customer to have the final findings presentation and the ADS submitted to him.

SPICE/Automotive SPICE does not define an assessment process, but only provides a relatively brief presentation of the requirements on SPICE-compliant assessments (see [Hoermann et al. 2006]). Based on this, organizations can define their own assessment processes. A working group within the VDA (German Association of the Automotive Industry [VDA]) is currently engaged in developing a standardized assessment process for the performance of supplier assessments. Documents such as assessment plan, final findings presentation, and assessment report are common in practice. In appraisals, the creation of a report is optional and not very common.

A similar construction can be found in CMMI: In the Appraisal Requirements for CMMI (ARC) basic requirements on appraisal methods[9] are distinguished. Three appraisal classes (A/B/C) are defined there, whereby A represents the most elaborate and most thorough method. Based on this, organizations may develop their own appraisal methods. SEI has defined its own appraisal method family called »SCAMPI«[10]. Associated training makes high demands on candidates, including participation in appraisals, several courses including numerous tests, and a practical examination (»observation«). There, candidates can prove their competence in their first appraisal as lead appraiser while being observed and assessed by an SEI-authorized observer. SEI distinguishes between two qualification levels: »SEI-authorized SCAMPI Lead Appraiser« (may perform SCAMPI A/B/C) and »SEI-authorized SCAMPI B&C Team Lead« (may perform SCAMPI B/C).

As regards SPICE/Automotive SPICE, the iNTACS (International Assessor Certification Scheme, see [iNTACS]) has emerged as a counterpart. iNTACS was recognized by HIS and VDA. Its training scheme was established in 2006 and currently lists over 30 member organizations, among them car manufacturers and suppliers, training and consultancy companies, as well as representatives from research institutions. Since its foundation, more than 400 assessors have been successfully trained according to a defined training scheme with independent examinations. More than 250 active assessors are currently registered worldwide. iNTACS distinguishes between provisional, competent and principal assessor[11]. To gain this qualification, a sound education, good theoretical model knowledge, practical assessment experience and continuous further professional training are required and checked. Training is considerably cheaper than the SEI training[12]. At the moment, most of the assessors registered at iNTACS come

---

9. Here »method« is used as a synonym for »process«.
10. Standard CMMI Appraisal Method for Process Improvement.
11. ISO/IEC 15504-3 only distinguishes between »provisional assessor« and »competent assessor«.
12. The reason for this is the SEI observation. As a rule, at least 30,000 USD need to be budgeted for an observed SCAMPI A, possibly with additional costs for simultanuous translation.

from German speaking countries, although the training system has been extended to numerous other European countries[13]. In addition, training has been performed in Japan, the United States, and Taiwan.

For comparison: On 12/31/2006, SEI had authorized 455 (SCAMPI V1.1) and 328 (SCAMPI V1.2) SCAMPI Lead Appraisers, as well as 456 SCAMPI B&C Team Leads from 54 countries (see [SEI Partner Directory]).

SCAMPI types A/B/C differ, among other things, in the depth and thoroughness of the appraisal, the size of the required team, and in the significance of the findings. »Ratings« and the award of »Levels« are exclusively reserved for SCAMPI A appraisals. The appraisal scope[14] is fairly variable in all SCAMPI methods; most SCAMPI A appraisals, however, are fairly broad in scope. For example, in order to identify a maturity Level 3 for a larger organization, a lot of effort is expended, involving a large number of projects and interviewees and an onsite examination period of several weeks[15]. The inventory of written evidence and the traceability of findings to the instances of evidence (written or verbal) are the pillars of each type of SCAMPI appraisal. Both are supported by the use of PII[16]-worksheets. Often, these are filled in by the organization with documented evidence many weeks before the appraisal[17] and completed during the interviews with further documented and particularly verbal evidence.

The term »PII« is not used in SPICE/Automotive SPICE; however, an equivalent concept is prescribed. Objective pieces of evidence collected for the purpose of process attribute evaluation, must be recorded. Traceability between the two must exist. Different »weight classes« similar to A/B/C are not defined in SPICE/Automotive SPICE, yet different types of assessment do exist. ISO/IEC 15504-3 refers to »self-assessments« and »independent assessments«. In practice, other types are common (see [Hoermann et al. 2006]). In the future, further details can be expected from ISO/IEC 15504, part 7 (Assessment of Organizational Maturity), which is currently under development.

---

13. Austria, Belgium, Bulgaria, Czech Republic, Denmark, England, France, Ireland, Italy, Lithuania, the Netherlands, Norway, Poland, Portugal, Rumania, Sweden, Switzerland, Slovenia, Spain, Turkey.
14. For instance, projects, support functions, staff to be interviewed.
15. With similar scope, a SPICE/Automotive SPICE assessment would require a similar amount of effort.
16. Practice Implementation Indicator.
17. Not inevitably, though, the method also allows the collection of evidence by the appraisal team itself (so-called »discovery mode«).

5 Functional Safety

Since more and more mechanical and hydraulic systems in safety related vehicle components are being replaced or at least complemented by electrical and electronic solutions, the question about the »functional safety« of these systems arises more and more. The customer must be able to rely on these systems and this issue is going to decisively influence his buying behavior. Dramatic declines in sales are well-known as a result of extensive public discussion of safety problems with chassis and brake systems in some car models. If in the case of product liability claims, it can be proven that the state of technology concerning development methods was not observed, manufacturers may be faced with high indemnity claims and legal costs, especially in the North American market.

The leading standard is [IEC 61508], which is designated as an »IEC basic safety publication«. This means it is to be used as a basis for application-specific standards. There is an ISO Working Group (ISO TC22 SC3 WG16) working on one of the standards based on IEC 61508 for safety functions in road vehicles (ASIL – Automotive Safety Integrity Level). Its aim is to prepare the ISO 26262 standard. Today, most car manufacturers already require their suppliers to develop control units and software in compliance with IEC 61508.

IEC 61508 is concerned with potentially hazardous failures or malfunctions that may have serious consequences, such as death or damage to the health of people or damage to the environment. Its aim is to recognize these hazards and to counteract them within maximum allowable statistical boundaries. In doing so, systems are identified that may contribute to such hazards. These systems are described as »safety related«. A single fault alone[1], or a fault in combination with other system faults, may constitute a hazard.

Based on a maximum-allowable upper limit for the probability of a potentially dangerous event occurring, »Safety Integrity Levels« (SIL) are demanded for all involved systems (in this case, for instance, sensors, cables, connectors, control units), depending on their degree of contribution: Four levels, from SIL 1 to SIL 4, are distinguished. Achievement of SIL 1 and 2 is considered relatively

---

1. This case is often called »safety-critical«, in combination with others as »safety related«, though the differences between the two terms are not clearly defined.

easy but already requires advanced design methods and thorough testing and reviews. The effort escalates drastically if SIL 3 is to be achieved; this is due, for instance, to the necessity of revalidation after changes. The same applies if SIL 4 is to be reached, as this requires the deployment of very elaborate methods such as formal design methods. The required SIL can be identified in two ways:

- Quantitatively: the frequency of hardware failures is predicted and compared with tolerable risk limits.
- Qualitatively: SIL is determined via qualitative evaluation, usually by means of a risk graph (see figure 5–1). This method is preferably used in case of software.

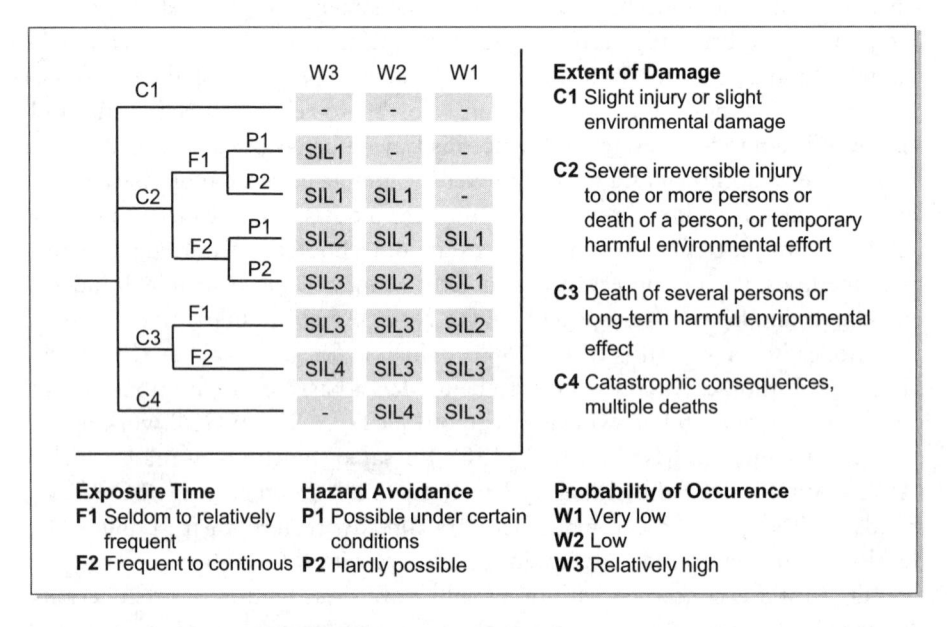

Figure 5–1 Example of a risk graph

IEC 61508 covers concept, planning, development, realization, installation, maintenance, and modification, up to the decommissioning of the system. In this context, planning, coordination, and documentation of all safety-related tasks are central management tasks that cover all phases. Regarding software development, these tasks include the following:

- Establish functional safety management, clearly define responsibilities
- Concept phase
    - Define safety related functions
    - Define software safety lifecycle for this function
    - Perform hazard analysis and risk evaluation
    - Establish a concept for functional safety

5 Functional Safety

- Development phase[2]
  - Plan software safety validation
  - A V-model like process model with specifications of software safety requirements, software architecture, software system design, module design, coding, unit test, integration test, validation
  - Planning of software safety-validation
  - Planning of software verification
  - Change management system with varying degrees of renewed verification and validation

The requirements for the development differ depending on the individual SIL level. This particularly concerns the, in some cases very detailed, prescriptions regarding methods to be applied, documentation, and configuration management (see figure 5–2).

Development Step	Method/Measure	SIL1	SIL2	SIL3	SIL4
Software architecture design	Structured design methods	++	++	++	++
Software design and software development	Certified compiler	+	++	++	++
Software design and software development: Support tools and coding languages	Library of proven in use/ verified modules and components	+	++	++	++
Software design and software development: Detailed design	Design and coding guidelines	+	++	++	++
Modification	FMEA	++	++	++	++
Modification	Repeated verification of changed software modules	++	++	++	++
Modification	Renewed verification of impacted software modules	+	++	++	++
Modification	Renewed validation of the complete system	o	+	++	++
Software verification	Formal proof	o	+	+	++
Software verification	Statistical tests	o	+	+	++
Legend: o = no recommendation for or against application; + = recommended; ++ = highly recommended					

Figure 5–2 *Examples for different SIL level requirements on development methods in the software development sector*

---

2. Here only shown for software. The standard describes measures for development at the system level as well as for hardware and software.

Implementation of IEC/61508 requires, among other things, mature development processes with associated methods and support tools. Many of the Automotive SPICE processes contribute to the accomplishment of these requirements. However, IEC/61508 requirements often clearly exceed the requirements of Automotive SPICE processes, especially regarding the higher SILs. There is no effective linkage of requirements in the standards (e.g., »level X in certain Automotive SPICE processes is equal to SIL Y«). To put it simply: The achievement of Level 2 in the HIS processes is a necessary (although not sufficient) prerequisite for the development of safety-critical software (SIL 1 and higher). Beyond that (i.e., Level 3/4/5) there is almost no generic practice that can be directly mapped onto one of the IEC/61508 requirements. The main difference is that Automotive SPICE requires *what* needs to be done while IEC/61508 goes on to require *how*, i.e., with which methods, this needs to be done.

A Overview of Selected Work Products

The following list is a selection of work products frequently used in practical application. References are made to processes for which examples or explanations are provided in this book.

Work Products		Process ID	Section
01-00	Configuration item	SUP.8	2.16
01-03	Software item	ENG.7	2.9
01-50	Integrated software	ENG.7	2.9
02-01	Commitment/agreement	ACQ.4	2.1
03-03	Benchmarking data	MAN.6	2.21
03-04	Customer satisfaction data	MAN.6	2.21
04-04	High level software design	ENG.5	2.7
04-05	Low level software design	ENG.5	2.7
04-06	System architecture design	ENG.3	2.5
07-01 - 07-08	Different exemplary metrics	MAN.6	2.21
08-04	Configuration management plan	SUP.8	2.16
08-12	Project plan	MAN.3	2.19
08-13	Quality plan	SUP.1	2.13
08-16	Release plan	SPL.2	2.2
08-17	Reuse plan	REU.2	2.23
08-27	Problem management plan	SUP.9	2.17
08-28	Change management plan	SUP.10	2.18
08-29	Improvement plan	PIM.3	2.22
08-50	Test specification	ENG.8	2.10
08-52	Test plan	ENG.8	2.10

Work Products		Process ID	Section
11-05	Software unit	ENG.6	2.8
11-06	System	ENG.9	2.11
13-01	Acceptance record	ACQ.4	2.1
13-07	Problem record	SUP.9	2.17
13-09	Meeting support record	ACQ.4	2.1
13-14	Progress status record	ACQ.4	2.1
13-16	Change request	SUP.10	2.18
13-17	Customer request	ACQ.4	2.1
13-18	Quality record	SUP.1	2.13
13-19	Review records	SUP.4	2.15
13-21	Change control record	SUP.10	2.18
13-22	Traceability record	ENG.2	2.4
13-25	Verification results	SUP.2	2.14
13-50	Test result	ENG.8	2.10
14-02	Corrective action register	SUP.1	2.13
14-06	Schedule	MAN.3	2.19
14-09	Work breakdown structure	MAN.3	2.19
15-01	Analysis report	SUP.9	2.17
15-05	Evaluation report	SUP.9	2.17
15-06	Project status report	MAN.3	2.19
15-07	Reuse evaluation report	REU.2	2.23
15-12	Problem status report	SUP.9	2.17
15-16	Improvement opportunity	PIM.3	2.22
16-03	Configuration management library	SUP.8	2.16
17-02	Build list	ENG.7	2.9
17-03	Customer requirements	ENG.1	2.3
17-08	Interface requirements	ENG.2	2.4
17-11	Software requirements	ENG.4	2.6
17-12	System requirements	ENG.2	2.4
19-10	Verification strategy	SUP.2	2.14

Glossary

A/B/C/D-Sample

The automotive industry works with so-called samples (A,B,C,D), providing components under development as prototypes with increasing functionality and integrating them in test vehicles. The sample phases are defined as follows (for details see section 2.2.2):

A-Sample: functional prototypes, usually with limited drivability and low degree of maturity

B-Sample: functional, basic prototypes with full drivability and a high level of maturity. They may be created using pilot tools.

C-Sample: fully functional sample manufactured with series production tools

D-Sample: same as C-Sample, provided by suppliers for the purpose of the design sample release

Acceptance

Acceptance is the written legal declaration that a work product (or service) has been accepted by the customer. The customer usually satisfies itself by means of tests or reviews that the work product has the specified features, behaves according to the specified requirements, and can be used for its intended purpose. Acceptance is an act typically performed jointly by customer and supplier.

Acquisition

In the ACQ processes (see section 2.1 in ACQ.4.), acquisition denotes the selection of suppliers and the control of products or product components produced by the selected suppliers.

Assessment

Evaluation of an an organization's process performance capability against a model (e.g., Automotive SPICE PAM). The goal is the rating and improvement of processes (process capability).

Assessment model

See Process assessment model

AUTOSAR
An initiative of leading OEMs and tier-one suppliers to develop a de facto standard for electric and electronic vehicle architectures (see also REU.2).

Base Practice
One or more activities performed as part of a process to generate the output work products. Base practices are indicators of the process dimension.

Baseline
A configuration describing a particular development level (e.g., a requirements baseline, design baseline, or baseline related to the production of releases). Associated and related configuration items are »preserved« when forming a baseline, i.e., they are protected against changes.

Black-Box test
This test compares the externally observable behavior at the external software interfaces (without knowledge of their structure) with the desired behavior. Black-Box tests are frequently equated with »functional tests«, although they can of course also include non-functional tests. *See also White-box test.*

Capability dimension
The capability dimension includes all the indicators needed to support the rating of the process attributes.

Capability level
Capability levels are a measure of process capability. The different capability levels are awarded depending on which practices or work products of a process assessment model can be evidenced. Automotive SPICE distinguishes between six maturity levels, from Level 0 »Incomplete« to Level 5 »Optimizing«.

Capability profile
Representation of the process ratings accomplished in the individual process attributes per process (using the N/P/L/F scale).

Code
Here, code signifies source code and the description of procedures and data structures in a programming language to enable further processing by a compiler.

Code coverage analysis
A method of analysis defining which software parts are covered by the test case suite and which are not. It allows an assessment as to whether or not additional test cases are necessary. Automotive SPICE uses the term »code coverage«; internationally it is known as »code coverage analysis« [BCS SIGIST].

Code inspection
In Automotive SPICE, this term is used as a synonym for »code review«. The term »inspection« is usually understood to denote a more formal type of review that follows a defined process performed by qualified reviewers and inspection leaders. Typically, the inspection process provides for the review object to be made available to reviewers some time prior to the meeting so that it can be inspected. During the inspection, the inspection results are then only briefly discussed, consolidated and recorded. *See also Code review.*

Code review

The application of a review methodology (*see Review*) on software. In its glossary, Automotive SPICE cites the [IEEE 610] definition: A meeting at which software is presented to project personnel, managers, users, customers, or other interested parties for comment or approval. *See also Code inspection.*

Commitment

A binding agreement or obligation voluntarily entered into by two or more parties.

Configuration

A group of configuration items that together represent a particular development status.

Configuration items

Configuration items (also referred to as CM elements) are work products (e.g., files, code files, documents) that are put under configuration management.

Configuration management

Process for the definition and management of configurations, allowing change control and change monitoring over a defined period. Configuration management allows access to individual configurations or configuration items (i.e., work products). Differences between individual configurations are readily identifiable. A configuration can be used to form a baseline; *see also Baseline.*

Configuration management system

A combination of one (or several) CM tool(s) (i.e., software supporting physical storage and administration) together with corresponding rules (instructions, processes, conventions regarding change management, versioning, access restrictions, etc.); also called CM system.

Customer

Automotive SPICE uses the term "customer" to describe the relationship between two partners, whereby one is the supplier and the other the customer of the respective development service. See the introductory pages of chapter 2 for a detailed description.

Customer requirements specification

A customer requirements specification describes in natural language and from a user perspective all the requirements, expectations and wishes pertaining to a planned work product, including all constraints.

Defined process

The process derived, documented, and adapted, if necessary (by means of »tailoring«), from a standard process, and implemented in the project or elsewhere in the organization.

Development environment

A group of tools and infrastructure that support the development processes, including planning tools, design tools, simulators, generators, editors, translators, debuggers, configuration management tools, as well as hardware like PCs, test environments, etc.

Dynamic analysis
The process of evaluating a system or component based on its behavior during execution [IEEE 610]. *See also Static analysis.*

Extendibility
The degree to which a software component can be changed, for instance, to enhance its functionality (see [ISO/IEC 9126] »Quality Attributes for Software«).

Flashing
In the automotive industry, flashing is understood as the programming of an EEPROM in an ECU, e.g., to upload a new software version.

Function list
Common tool in the automotive industry used for planning the contents of individual samples and SW releases. The function list assigns functions (perhaps also bugfixes) to different SW releases and milestones and is used to monitor their implementation. It also supports early agreement between supplier and OEM regarding the functional scope.

Functional requirement
A requirement that directly influences and describes functionality.

Generic practice
One or more activities supporting the performance of a process from the perspective of the associated process attribute. Generic practices are indicators of the capability level dimension.

Hardware
Collective term for components, component assemblies, devices, and equipment used for data processing and computing.

Indicator
An indicator is an objective feature or attribute supporting the implementation of a process. Indicators are used for the rating of process attributes. The indicators of the process dimension are base practices and work products; the indicators of the capability dimension are generic practices and generic resources.

Integration
Stepwise assembly of components to create a product, usually accompanied by a variety of tests.

Lessons learned
The systematic capturing of experiences to learn for the future and to avoid mistakes. Lessons learned sessions can, for instance, be held at completion of a project or project phase.

Lifecycle model
The lifecycle model is a suitable and adequate approach to structure a project. It consists of the project phases that divide the project into larger sections and the description of the phase transitions (e.g., quality gates). If needed, it also defines subordinate activities and their interrelationships, e.g., in the form of loops and iterations.

Glossary

Measurement
Used in the context of the MAN.6 process. Measurement is understood as a continuous process during which process metrics are defined and measurement data are collected, analyzed, and evaluated. The objective is to understand, control, and optimize processes, for instance, to improve project control, reduce development effort and cost, or to improve on work products.

Mechanics
Within this book we understand mechanics as a collective term comprising mechanical parts needed to build engines, machines, devices, and tools. We delimit this term from »hardware« (see glossary). Examples of mechanical parts are casings, bearings, rollers, frames, etc.

Metric
Also called »measurement«. A numerical value describing a procedure, process, product attribute, or goal. A distinction is made between basic metrics (that can be measured directly) and derived metrics which result from mathematical operations using basic metrics.

Milestone
A milestone is an important event in the project at which significant work results have reached a particular development level, e. g., completion of a project phase or delivery of a sample version at a planned date.

MISRA
In the automotive industry, MISRA-C is a C programming standard developed by MISRA (The Motor Industry Software Reliability Association).

Modularity
Degree to which a software system consists of self-contained, small, logical units (software units) which interact with each other. The complexity of the interfaces between these software units must be manageable.

Non-functional requirements
Non-functional requirements have no direct impact on functionality. As regards software, non-functional requirements include complexity, level of nesting, testability, and maintainability. Non-functional requirements may, for example, result from the operational environment, e.g., requirements regarding a temperature range in which a system needs to work correctly (e.g., –58 to 176 degrees Fahrenheit).

Platform
In the automotive industry, a platform describes a technical basis on which to build models that appear outwardly different. In electronic development, a platform is understood as a hardware, software, or system construction kit that allows the easy creation of derivates via modification, parameterization, or derivation. All essential features, however, are preserved.

Process
A process consists of a sequence of (possibly parallel or alternative) activities or steps that are performed for a particular purpose and which transfer input work products into output work products. The output work products may be used by subsequent processes.

Process assessment model
A process assessment model (PAM) contains all the details for the assessment of process capability (so-called indicators) and is organized in two dimensions (process dimension and capability dimension). A process assessment model refers to one or several process reference models. One example of a process assessment model is the Automotive SPICE PAM.

Process attribute
Process attributes are attributes whose fulfilment can be rated. They are used for the assessment of the capability level achieved by a process. They are applicable to all processes.

Process dimension
The process dimension contains indicators for all relevant processes regarding process purpose and process outcomes. Processes are organized in process groups.

Process outcomes
The available and verifiable result of the successful implementation of a process.

Process owner
The process owner (person or team) is responsible for process definition and maintenance. At the organizational level, the process owner is responsible for the description of the standard process. At the project level, the process owner is responsible for the defined process. A process may therefore have several process owners with varying levels of responsibility.

Process profile
See Capability profile

Process Reference Model
A Process Reference Model (PRM) describes a set of processes associated with a particular application domain. Each process is described in terms of process purpose and necessary process outcomes to accomplish the process purpose. One example of a process reference model is the Automotive SPICE PRM.

Project
A temporary, one-time endeavour undertaken to create a product or service.

Project attributes
Project attributes like business and quality goals, project scope and complexity, effort, schedule, and budget.

Project management
Planning, monitoring, and control of a project with the objective to deliver a product which meets the agreed requirements regarding product/features, quality, schedule, and costs.

Project phases

Project phases consist of chronologically and logically associated, comprehensive project activities. Examples of project phases are project start, project planning phase, project execution phase, project completion. *See also Lifecycle model.*

Project plan

The project plan consists of one or several planning documents (e.g., work breakdown structure, schedule, resource planning) that define the project scope and essential project attributes. It may also consist of a directory structure with different files. The project plan is the basis for project control. If the project plan consists of several planning documents, care must be taken that, in sum, the individual documents represent a conclusive, coherent whole.

Quality assurance

According to ISO 9000:2000 [ISO 9000], paragraph 3.2.11, quality assurance is defined as »part of quality management focused on providing confidence that quality requirements will be fulfilled«. Automotive SPICE describes a quality assurance process particularly for software development, whose activities aim at the fulfillment of quality requirements.

Rating scale

According to ISO/IEC 15504, process attributes are rated based on a 4-stage rating scale (N/P/L/F) (see section 3.2 for a description of the scale).

Regression testing

Testing to verify that previously successfully tested features are still correct. It is necessary after modifications to eliminate undesired side effects.

Release

Formal decision that is based on project results regarding the maturity level of a work product and taken for the work product's official delivery to the next process step or customer.

Consistent set of versioned objects with defined features and attributes, intended for delivery to an internal or external customer. *See also Version.*

Release planning

Planning which defines in which versions particular features of a product are implemented. Based on this, the development process can be structured, and work can be prioritized accordingly.

Reliability

Reliability describes a product's ability to maintain its defined functions under defined conditions for a specified period of time.

Requirement

A requirement describes a verifiable feature or service of a product, system, or process that is to be satisfied.

Reuse

Use of existing products or product components in other applications.

Review
: Formal check of an object (e.g., document or code) against specifications and applicable guidelines, performed by reviewers. The aim is to identify defects, weaknesses, or gaps of the review object, to comment, document, and determine the object's expected capability level.

Risk
: A risk is an undesired event or potential problem which may occur with a certain probability sometime in the future. Risk occurrence is associated with damage; i.e., it has a negative effect on project goals. It may cause cost increases, schedule shifts, quality problems, or other damages.

Risk management
: Risk management is a continuous process to be performed throughout the entire life of a project, and an important part of project management activities. The objective of risk management is to identify and prevent risks, to reduce their probability of occurrence, or to mitigate the effects in case of risk occurrence.

Schedule
: The schedule contains all the activities that need to be performed, including their sequence and interdependencies, milestones, duration, estimated effort, start and end dates, as well as resource assignment. In addition, it should detail the critical path or paths.

SIL
: Classification of safety-critical systems using Safety Integrity Levels (SIL) (see also chapter 5).

Software item
: Source code, object code, job control code, control data, or a collection of these items (according to [IEEE 610]). In some cases, Automotive SPICE uses »software components« as a synonym.

Software or system requirements specification
: These are the contractually binding, detailed descriptions of a service to be rendered, for example of a planned appliance, technical installation, machine, tool, or software program. In contrast to the customer requirements specification, its contents are precise, complete, traceable, and associated with technical specifications which also define the operational and maintenance environment. They include the interpretation of requirements on work products and services from a developer's perspective, including conditions and constraints. They also include a detailed solution concept for the implementation of the customer requirements.

Software test
: Verification or validation of a program, *see also Testing*.

Software unit
: A software item that cannot be further subdivided into smaller components. The term used in Automotive SPICE is »unit« or »software unit« (see [IEEE 610]).

Glossary

Sponsor
Within the context of an assessment, a sponsor is someone who supports the assessment by providing human and material resources, capital, and services.

Stakeholders
Individuals who supply inputs to activities and who are either involved in or affected by them. Also, people affected by or using the results of such activities, including people who provide resources.

Standard process
A standardized process that is applied across a particular section of the development organization. A standard process consists of fundamental process elements, such as process activities with their dependencies and interfaces, input and output work products, support tools, and facilities. It also includes information on which roles are involved in the activities.

Static analysis
Analysis of a program carried out without executing the program [BCS SIGIST]. *See also Dynamic analysis.*

Steering Committee
Board in which several people (for instance, management representatives, representatives of different customer interest groups) constitute the highest supervisory body of a project.

Strategy
In Automotive SPICE, the term strategy is frequently used in connection with different processes (e.g., CM strategy, validation strategy). Usually, a strategy is asked for in the first base practice of a process, requiring the definition of the process's general course of action, i.e., its activities and the times when they are to be performed.

Supplier
Automotive SPICE uses the term »supplier« for the description of the relationship between two partners, in which one is the supplier and the other the customer of the (development) service. In the context of a contractual relationship, the term »supplier« denotes the contractor, in a call for bids the »supplier« is the vendor.

System
Product or product component that in turn consists of interacting subsystems.

Test case
A combination of input data, conditions, and expected output data developed for the functional or non-functional verification of a test object to see whether it comforms with a specified and agreed requirement.

Test plan
A document describing the scope, approach, resources, and schedule of intended testing activities (see [IEEE 829]).

Testing
Activity to verify if an object conforms with its requirements and to detect deviations.

Tier
Used to describe the level or layer in the supplier chain. Suppliers are divided into tier one, tier two suppliers, etc. In the automotive industry there often exists a complete hierarchy of customer-supplier relationships. For instance, a tier one supplier acquires additional system components from his own subcontractor, or from subcontractors stipulated by the customer (»tier two«).

Timing behavior
Ability of a software product to ensure adequate response and process times under defined conditions.

Traceability
Starting from requirements, traceability establishes a correlation between elements of different development steps (see section 2.24).

Traceability matrix
A matrix describing the traceability between requirements and work products.

Unit test
Testing of the software units.

Use case diagram
Use case diagrams are used to illustrate required functionality graphically and textually. They are intuitively understood and may serve as a basis for discussions between customer and supplier. Each use case diagram consists of use cases (ellipses) and actors (stick man figures) that are in a closed system. A use case is an activity typically comprised of a noun and a verb and comprising several activities (e.g., system start-up, press button, read display). A use case diagram is often used in the context of a software requirements analysis and is a component of the Unified Modeling Language (UML).

Validation
Validation answers the question »Am I building the right system?«, i.e., is it appropriate for its intended use? Hence, validation checks whether a system is suited for its purpose. This is primarily done by checking against the customer and system requirements.

Verification
Verification denotes a phase-specific process to evaluate whether the work products of a development phase are correct and complete with regard to their immediate requirements. For instance, during verification of a software component particular focus is on checking against design specifications and coding guidelines. Verification answers the question »Am I building the system right?«, i.e., does it meet the requirements?

Version
Uniquely defined object representing a defined development level (»snapshot«). It is distinguished from other versions by a unique name and typically managed by configuration management.

White-box test
This test is derived knowing the inner structure of the software and based on the program code, design, interface descriptions, and so on. White-box tests are also called» structure based tests«. *See also Black-box test.*

Work Breakdown Structure
A work breakdown structure (WBS) is an arrangement of project elements consisting of deliverables or project phases. It structures and defines the overall project content and scope.

Work Product
An artifact associated with the performance of a process. Output work products are created or modified during a process. After its completion, they are available to subsequent processes. Work products can be project-internal items or be delivered externally. In the latter case the term »product« is usually used. Output work products are indicators of the process dimension.

Abbreviations

AUTOSIG	Automotive Special Interest Group
BP	Base practice
CCB	Change Control Board
CM	Configuration management
CMM	Capability Maturity Model
CMMI	Capability Maturity Model Integration
COTS	Commercial off-the-shelf
CRB	Change Request Board
ECU	Electronic control unit
EFQM	European Foundation for Quality Management
EITVOX	Entry criteria, Inputs, Tasks, Verification, Outputs, eXit Criteria
ETA	Event tree analysis
FAKRA	Automotive Standards Committee in the German Institute for Standardization
FMEA	Failure Modes and Effects Analysis
FTA	Fault tree analysis
GAMP	Good automated manufacturing practice
GP	Generic practice
GQM	Goal Question Metric
GR	Generic resource
HAZOP	Hazard and Operability Study
HIS	Working group established by Audi, BMW, Daimler, Porsche, and Volkswagen
IEC	International Electrotechnical Commission
IEEE	Institute of Electrical and Electronics Engineers

iNTACS	»International Assessor Certification Scheme«, see *www.intacs.info*
IS	International Standard
ISO	International Organisation for Standardization
iSQI	International Software Quality Institute, see *www.isqi.org*
MISRA	Motor Industry Software Reliability Association
OEM	Original equipment manufacturer, in this case the vehicle manufacturer
OIL	Open issues list
PA	Process attribute
PAM	Process assessment model
PM	Project management
PRM	Process reference model
QA	Quality assurance
RCA	Root cause analysis
SCAMPI	Standard CMMI Appraisal Method for Process Improvement
SEI	The Carnegie Mellon® Software Engineering Institute, Carnegie Mellon University, Pittsburgh, USA (publishers of CMM and CMMI)
SEPG	Software engineering process group
SIL	Safety integrity level
SOP	Start of production
SPICE	Software Process Improvement and Capability Determination
SW	Software
TR	Technical report
UML	Unified Modeling Language
VDA	Association of the Automotive Industry
WBS	Work breakdown structure

Literature, Standards, and Web Pages

[Ahren et al. 2001]
　　Ahren, D.; Clouse, A.; Turner, R.: CMMI Distilled. Addison-Wesley, Boston, 2001.

[Automotive SPICE]
　　www.automotivespice.com
　　Download of the Automotive SPICE™Process Reference Model (PRM) and Process Assessment Model (PAM) is free upon registration.

[AUTOSAR]
　　www.autosar.org
　　Homepage of AUTOSAR (Automotive Open System Architecture).

[BCS SIGIST]
　　www.sigist.org.uk bzw. *www.testingstandards.co.uk/living_glossary.htm*
　　Working Draft: Glossary of terms used in software testing, Version 6.3
　　Produced by the British Computer Society, Specialist Interest Group in Software Testing (BCS SIGIST)

[Chrissis et al. 2003]
　　Chrissis, M.; Konrad, M.; Shrum, S.: CMMI—Guidelines for Process Integration and Product Improvement. Addison-Wesley, Boston, 2003.

[CMM 1993a]
　　Paulk, M.; Curtis, B.; Chrissis, M.; Weber, C.: Capability Maturity Model for Software, Version 1.1, Technical Report CMU/SEI-93-TR-024. Software Engineering Institute, Carnegie Mellon University, 1993.

[CMM 1993b]
　　Paulk, M.; Weber, C.; Garcia, S.; Chrissis, M.; Bush, M.: Key practices of the Capability Maturity Model, Version 1.1, Technical Report CMU/SEI-93-TR-025. Software Engineering Institute, Carnegie Mellon University, 1993.

[CMMI 2006]
　　Capability Maturity Model® Integration (CMMI[SM]), for Development, Version 1.2—Technical Report CMU/SEI-2006-TR-008. Software Engineering Institute, Carnegie Mellon University, August 2006.

[ISO 9000]
: ISO 9000:2005: Quality management systems—Fundamentals and vocabulary

[ISO 9001]
: ISO 9001:2000: Quality management systems—Requirements

[DoD 1998]
: Software Acquisition Best Practices Initiative. Department of Defense, Computer & Concepts Associates, USA, 1998.

[Fenton et al. 1997]
: Fenton, N. E.; Pfleeger, S. L.: Software Metrics—a rigorous & practical approach. 2nd ed., PWS Publishing Company, 1997.

[Freedman et al. 1990]
: Freedman, D.P.; Weinberg, G.M.: Handbook of Walkthroughs, Inspections, and Technical Reviews. ISBN 0-932633-19-6

[Gilb 1993]
: Gilb, T.: Software Inspection. Addison-Wesley, 1993.

[Hansen]
: *www.hansenreport.com*
Monthly published Business and Technology Newsletter; The Hansen Report on Automotive Electronics, 150 Pinechurst Road, Portsmouth NH 03801

[Hindel et al. 2006]
: Hindel, B.; Hoermann, K.; Mueller, M.; Schmied, J.: Basiswissen Software-Projektmanagement. 2. Auflage, dpunkt.verlag, Heidelberg, 2006.

[HIS]
: *www.automotive-his.de*
Website providing further information on the OEM Initiative Software (HIS). The latest HIS Scope can be downloaded via the link »Process Assessment« listed under Working Groups.

[Hoermann et al. 2006]
: Hoermann, K.; Dittmann, L.; Hindel, B.; Mueller, M.: SPICE in der Praxis—Interpretationshilfe für Anwender und Assessoren. dpunkt.verlag, Heidelberg, 2006.

[IEC 61508]
: IEC 61508, Functional safety of electrical/electronic/programmable electronic safety-related systems, International Electrotechnical Commission.
This standard consists of seven parts ratified as an »International Standard« between 1998 and 2000, and one »Technical Report«, published in 2005. See *www.iec.ch/zone/fsafety/* for details.

[IEEE 610]
: IEEE Std 610.12-1990 (R2002), IEEE Standard Glossary of Software Engineering Terminology, IEEE Computer Society.

[IEEE 829]
IEEE Std 829-1998, IEEE Standard for Software Test Documentation, IEEE Computer Society.

[IEEE 830]
IEEE Std 830-1998, IEEE Recommended Practice for Software Requirements Specifications, IEEE Computer Society.

[IEEE 1012]
IEEE Std 1012-1998, IEEE Standard for Software Verification and Validation, IEEE Computer Society.

[IEEE 1028]
IEEE Std 1028-1997, IEEE Standard for Software Reviews, IEEE Computer Society.

[iNTACS]
www.intacs.info
Website of the International Assessor Certification Scheme, officially recognized by the OEM Initiative Software (HIS) for assessor certification. The site offers many useful iNTACS documents for download

[ISO/IEC 12207]
- ISO/IEC 12207, (1995-08) Information technology–Software Lifecycle Processes;.
- ISO/IEC 12207 AMD 1 (2002-05) Information technology—Software Lifecycle Processes; Amendment 1.
- ISO/IEC 12207 AMD 2, (2004-11) Information technology—Software Lifecycle Processes; Amendment 2.

[ISO/IEC 15504]
2003-2006
- ISO/IEC 15504-1, Information technology—Process assessment—Part 1: Concepts and vocabulary.
- ISO/IEC 15504-2, Information technology—Process assessment—Part 2: Performing an assessment.
- ISO/IEC 15504-3, Information technology—Process assessment—Part 3: Guidance on performing an assessment.
- ISO/IEC 15504-4, Information technology—Process assessment—Part 4: Guidance on use for process improvement and process capability determination.
- ISO/IEC 15504-5, Information technology—Process assessment—Part 5: An exemplary Process Assessment Model.

[ISO/IEC 9126]
ISO/IEC 9126-1:2001, Software engineering—Product Quality—Part 1: Quality model.

[Kerzner 2001]
Kerzner, H.: Project Management. 7th ed., Wiley, 2001.

[Kneuper 2006]
Kneuper, R.: CMMI—Verbesserung von Softwareprozessen mit Capability Maturity Model Integration. 2. Auflage, dpunkt.verlag, Heidelberg, 2006.

[MISRA]
www.misra.org.uk
The Motor Industry Software Reliability Association has published C coding guidelines, especially for C developers in the automotive industry. Also refer to the subset provided by HIS (*www.automotive-his.de*).

[Mueller 2004]
Mueller, M.: Project Support & Control Office (PSO™): Metric-based Project Management, Presentation on ESEPG, London, 2004.

[Phillips 2006]
Phillips, M.: CMMI Version 1.2 and Beyond, NDIA 6th Annual CMMI Technology Conference, Denver, November 13–16, 2006.

[PMBOK 2004]
PMBOK: A guide to the Project Management Body of Knowledge, 2004 Edition, Project Management Institute, 2004.

[PPSM 1998]
Process Professional Supplier Management Part 1-7, United Kingdom defense evaluation research agency, Process professional library services, UK, 1998.

[Published Appraisal Results]
http://sas.sei.cmu.edu/pars/
CMMI (SCAMPI)Appraisal results voluntarily published by assessed organizations

[Radice et al. 1988]
Radice, Ronald A., Phillips, Richard W.: Software Engineering, An Industrial Approach. Prentice Hall, Englewood Cliffs, New Jersey, 1988.

[SA-CMM 1999]
Cooper, J.; Fisher, M.; Sherer, S. W. (eds.): Software Acquisition Capability Maturity Model (SA-CMM), Version 1.02, CMU/SEI-99-TR-002. Carnegie Mellon, Software Engineering Institute 1999.

[SEI]
www.sei.cmu.edu
Website of the SEI (Software Engineering Institute), offering a wealth of information, technical reports, etc., on CMM, CMMI, and topics related to software and systems engineering

[SEI Partner Directory]
http://partner-directory.sei.cmu.edu/
Site supporting the search for SEI partners and SEI-authorized individuals

[SEI Rep]
> *http://seir.sei.cmu.edu*
> Software Engineering Information Repository of the SEI (Software Engineering Institute), providing examples and documentation related to capability maturity models (particularly CMMI) and different processes

[Spillner et al. 2005]
> Spillner, A.; Linz, T.: Software Testing Foundations—A Study Guide for the Certified Tester Exam. 2nd edition, Rocky Nook, Santa Barbara, CA, 2007.

[van Solingen et al. 1999]
> van Solingen, R.; Berghout, E.: The Goal/Question/Metric Method. McGraw-Hill, 1999.

[V-Modell]
> *www.v-modell.iabg.de*
> Site offering free download of the German V-Model.

[V-Modell XT]
> *www.v-modell-xt.de*
> Website of the V-Model XT.

[VDA]
> *www.vda.de*
> German Association of the Automotive Industry (VDA), based in Frankfurt/Main, Germany. The site offers information on Automotive SPICE assessor training.

Index

A

activities of a project 165
analysis 42
    code coverage 72
    dynamic 68
    risk 36
    static 68
    system requirements 38, 40, 42
appraisal 260
A-sample 24
assessment 104
Audit 100
Automotive Special Interest Group (AUTOSIG) 3, 39
AUTOSAR 201
AUTOSIG *see Automotive Special Interest Group*

B

baseline 40, 127, 131, 133
    audit 134
benchmarking 190
black-box test 89
boundary value analysis 89
branch 131
    coverage 89
B-Sample 161
B-sample 24
build 27

C

capability
    dimension 8
    level 8, 213, 216
Capability Maturity Model Integration (CMMI) 253, 255
CCB *see Change Control Board*
Change Control Board (CCB) 153
change management 34
Change Request Board (CRB) 44, 153
CMMI *see Capability Maturity Model Integration*
Cockpit Chart 189
code
    coverage analysis 72
    inspection 68
    review 68
coding guideline 52
commitment 194
communication plan 228, 229
competence 240, 246
condition coverage 89
configuration management system 34
continuous process improvement 251
continuous representation 255
control limit 249
corrective measure 169
cost 162
    estimation 162
CRB *see Change Request Board*
C-sample 24

customer 13
    requirements 37, 38, 40, 44, 45
C0 metric 89
C1 metric 89

D

data
    collection 186
    historical 163
    measurement 182
defect injection 146
defined process 234, 236
dependencies 232
domain 202
DOORS 40
D-sample 25
dynamic analysis 68

E

ECU 16
effort estimation 162
EFQM 101
equivalence class 89
estimate 163
estimation
    cost 162
    effort 162
    expert 164
    schedule 162
event tree analysis 174
expert estimation 164

F

Failure Modes and Effects Analysis 174
failure scenarios 58
fault tree analysis 174
feedback meeting 188
FMEA *see Failure Modes and Effects Analysis*
formal review 120
framework 213
Fully achieved 10, 215
function 24
    list 159
functional requirement 40
functionality increase 187

G

Generic practices (GP) 214
Generic resources (GR) 214
Goal/Question/Metric method (GQM method) 182, 183
GP *see Generic practices*
GQM method *see Goal/Question/Metric method*
GR *see Generic resources*
gray-box test 89

H

Hazard and Operability Study (HAZOP) 42, 54
HAZOP *see Hazard and Operability Study*
HIS *see OEM software initiative*
historical data 163
horizontal traceability 82, 88, 94, 98, 205, 206, 208, 211

I

IEC 61508 263
improvement objectives 197
infrastructure 202
inspection 120
iNTACS *see International Assessor Certification Scheme*
interface 49
    requirement 45
International Assessor Certification Scheme (iNTACS) 261
ISO Working Group 263
ISO 26262 263
ISO 9001 101
ISO/TS 16949 101

J

joint review 100
joint technical review 19

L

label 132
Largely achieved 10, 215
Lessons Learned 163

Index

M

management
    commitment 194
    review 120, 243
maturity models 1
measurement 182
    concept 184
    data 182, 187
    framework 213
    program 182
metric 185, 191
milestone 161
    review 120
MISRA *see Motor Industry Software Reliability Association*
Motor Industry Software Reliability Association (MISRA) 52

N

nonfunctional requirement 40, 52, 69
Not achieved 10, 214
N/P/L/F scale 10, 214

O

OEM software initiative (HIS) 3
OIL *see open issues list*
open issues list (OIL) 20
operating environment 54
organizational standard 234

P

Partially achieved 10, 214
PII worksheet *see practice implementation indicator worksheet*
practice implementation indicator (PII) worksheet 262
prediction accuracy 249
probability of occurrence 177
problem
    cause 199
    class 122
    record 110
    state 139

process
    activity 223
    assessment model 7
    attribute 213
        rating 10, 214
    capability 269
    capability indicators 214
    capability level model 215
    compliance audit 106
    defined 234, 236
    dimension 7, 11
    documented 220
    elements 236
    group 195
    improvement, continuous 251
    interpretation 11
    purpose 217
    reference model 6
    standard 234
progress monitoring 20, 167
project 156
    cockpit chart 189
    completion review 163
    meeting 168
    phase 160
    scope 157
purpose of the process 217

Q

quality
    assurance 99
        plan 103, 110
        strategy 103
    management 100
    plan 110
    planning 162
    record 111
    report 108
    requirements 52
quantitative understanding 249

R

rating scale 10, 214
RCA *see root cause analysis*
release 24
   criteria 28
   documentation 28, 29
   note 29
   planning 25, 43, 55
representation
   continuous 255
   staged 255
requirement
   customer 37, 38, 40, 44, 45
   functional 40
   interface 45
   nonfunctional 40, 49, 52, 69
   software 53, 57, 59, 63
   system 40, 44, 49
reuse 200
review 119
   criterion 121
   formal 120
   joint 100
   log 126
   management 120, 243
   milestone 120
   object 123
   planning 121
   process 122
   result 122, 124
   technical 120
risk 173
   analysis 36, 41
   checklist 181
   identification 177
   indices 177
   management strategy 175
   priority number 177
   tracking 179
role description 240
root cause analysis (RCA) 144

S

Safety Integrity Level (SIL) 263
sample 24
SCAMPI *see Standard CMMI Appraisal Method for Process Improvement*
schedule estimation 162
scope statement 157
SEI *see Software Engineering Institute*
SEPG *see Software Engineering Process Group*
SIL *see Safety Integrity Level*
SIL test *see software-in-the-loop test*
software
   integration 92
   requirement 53
   requirements specification 53
Software Engineering Institute (SEI) 260
Software Engineering Process Group (SEPG) 107
software requirements 57, 59, 63
software-in-the-loop test (SIL test) 76
SOP *see start of production*
staged representation 255
stakeholder 166
Standard CMMI Appraisal Method for Process Improvement (SCAMPI) 260
standard process 234
start of production (SOP) 156, 184
statement coverage 89
static analysis 68, 89
supplier 13, 16
   management 16
system 39, 41
   architecture 46, 48, 49
   requirement 40, 41, 44, 49
   requirements analysis 38, 40, 42
   supplier 16
system requirements 40, 49
   analysis 42
   nonfunctional 49

T

tailoring 235
technical review 120
test
    black-box 89
    case specification 74
    design specification 74
    gray-box 89
    incident report 74
    item transmittal report 74
    log 74
    plan 42, 74
    procedure specification 74
    summary report 75
    white-box 89
testability 42
tier one 16
traceability 44, 45, 50, 56
    horizontal 72, 82, 88, 94, 98
    matrix 45, 59, 206
    vertical 63, 71, 205

U

understanding, quantitative 249
Unified Modeling Language (UML) 52
unit test 67
use case diagram 52

V

variant generation 235
verification 67, 100
    criterion 44, 69
    strategy 66
vertical traceability 63, 71, 205
V-model 161

W

WBS *see work breakdown structure*
white-box test 89
work
    breakdown structure 157
    package 164
    product overview 267

KUGLER MAAG CIE

YOUR WORLDWIDE PARTNER

KUGLER MAAG CIE is an international consulting company working in industries like automotive, finance, IT, railway and health care. KUGLER MAAG CIE specializes in systematic and sustainable improvements of products and services.

Based on industry standards like CMMI®, SPICE (ISO/ IEC 15504), Automotive SPICE™, Safety (IEC 61508, ISO 26262) we help our customers to realize measurable improvements in budget, quality, and time. Our employees are experts with significant experience in the implementation of improvement programs.

For SPICE-based services we have more than 20 SPICE Assessors at your disposal. Among them are leading experts with more than 12 years of experience in hundreds of assessments, three SPICE Principal Assessors (iNTACS™) and trainers who have qualified over 200 assessors.

KUGLER MAAG CIE supports the VDA (German Automotive Association) in the implementation of Automotive SPICE™ worldwide, is a founding member of iNTACS™ as well as a partner of the Software Engineering Institute (SEI), USA, sponsor of SEI-Europe and member of Lero, the Irish Software Engineering Research Centre.

KUGLER MAAG CIE GmbH
Leibnizstr. 11
70806 Kornwestheim, Germany

Phone: +49 7154 - 807 210
Fax: +49 7154 - 807 229

Internet: www.kuglermaag.com
E-Mail: information@kuglermaag.com

intacs.info
International Assessor Certification Scheme™

The International Assessors Certification Scheme, iNTACS™, has been developed in response to industry needs to qualify and train Assessors regarding principles and practices of Process Assessments on the basis of ISO/IEC 15504 and related standards.

iNTACS™ supports the Automotive Association VDA, the International Software Quality Institute, iSQI, and Training Provider in Accreditation and Training of the standards ISO/IEC 15504 and Automotive SPICE® worldwide.

iNTACS Office
Breite Straße 2d
D-14467 Potsdam

Phone: +49 (0) 331 231 810-80
Fax: +49 (0) 331 231 810-81
Email: office@intacs.info

www.intacs.info